Lecture Notes in Networks and Systems

Volume 180

The series "Lecture Notes in Networks and Systems" publishes the latest developments in Networks and Systems—quickly, informally and with high quality. Original research reported in proceedings and post-proceedings represents the core of LNNS.

Volumes published in LNNS embrace all aspects and subfields of, as well as new challenges in, Networks and Systems.

The series contains proceedings and edited volumes in systems and networks, spanning the areas of Cyber-Physical Systems, Autonomous Systems, Sensor Networks, Control Systems, Energy Systems, Automotive Systems, Biological Systems, Vehicular Networking and Connected Vehicles, Aerospace Systems, Automation, Manufacturing, Smart Grids, Nonlinear Systems, Power Systems, Robotics, Social Systems, Economic Systems and other. Of particular value to both the contributors and the readership are the short publication timeframe and the world-wide distribution and exposure which enable both a wide and rapid dissemination of research output.

The series covers the theory, applications, and perspectives on the state of the art and future developments relevant to systems and networks, decision making, control, complex processes and related areas, as embedded in the fields of interdisciplinary and applied sciences, engineering, computer science, physics, economics, social, and life sciences, as well as the paradigms and methodologies behind them.

Indexed by SCOPUS, INSPEC, WTI Frankfurt eG, zbMATH, SCImago.

All books published in the series are submitted for consideration in Web of Science.

More information about this series at http://www.springer.com/series/15179

Ripon Patgiri · Sivaji Bandyopadhyay ·
Valentina Emilia Balas
Editors

Proceedings of International Conference on Big Data, Machine Learning and Applications

BigDML 2019

Springer

Editors
Ripon Patgiri
Department of Computer Science and Engineering
National Institute Of Technology Silchar
Silchar, Cachar, Assam, India

Sivaji Bandyopadhyay
CSE Department
National Institute of Technology Silchar
Silchar, Cachar, India

Valentina Emilia Balas
Department of Automatics and Applied Software
Aurel Vlaicu University of Arad
Arad, Romania

ISSN 2367-3370 ISSN 2367-3389 (electronic)
Lecture Notes in Networks and Systems
ISBN 978-981-33-4787-8 ISBN 978-981-33-4788-5 (eBook)
https://doi.org/10.1007/978-981-33-4788-5

This Springer imprint is published by the registered company Springer Nature Singapore Pte Ltd.
The registered company address is: 152 Beach Road, #21-01/04 Gateway East, Singapore 189721, Singapore

Preface

International Conference on Big Data, Machine Learning, and Applications (BigDML 2019) focuses on both theory and applications in the broad areas of Big Data and Machine Learning. This conference aims to bring together the academia, researchers, developers, and practitioners from scientific organizations and industry to share and disseminate recent research findings in the fields of Big Data, Machine Learning and its applications. BigDML is an outstanding platform to discuss the key findings, exchanging novel ideas, listening to the world-class leaders and sharing experiences with peer groups. The conference provides the opportunities for collaboration with national and international organizations of repute to the research community. BigDML has a large number of participants and submissions from worldwide.

There are 152 submissions and BigDML issues 32 acceptance. The selected 21 papers will be published in this conference proceedings. Apart from that, there are world-leading keynote speakers, namely, Padma Shri Profs. Ajay Kumar Ray, Josef Van Genabith, Paolo Rosso, Alexander Gelbukh, Punam Kumar Saha, and Alain Tremeau. Finally, the conference concluded with a big success.

Silchar, India — Ripon Patgiri
Cachar, India — Sivaji Bandyopadhyay
Arad, Romania — Valentina Emilia Balas

Contents

Editors and Contributors

About the Editors

Dr. Ripon Patgiri is Assistant Professor at the Department of Computer Science & Engineering, National Institute of Technology Silchar. He has received his Ph.D. degree from National Institute of Technology Silchar. He has seven years of teaching and research experiences. Moreover, he has rich experiences in organizing conferences. He has published several journal articles, conference papers and book chapters. Also, he is editing several books. He is a senior member of IEEE.

Prof. Sivaji Bandyopadhyay is Director of National Institute of Technology Silchar since December 2017. He is Professor of the Department of Computer Science & Engineering, Jadavpur University, India, where he has been serving since 1989. He is attached as Professor, Computer Science and Engineering Department, National Institute of Technology Silchar. He has more than 300 publications in reputed journals and conferences. He has edited two books so far. His research interests are in the area of natural language processing, machine translation, sentiment analysis and medical imaging among others. He has organized several conferences and has been the Program Committee member and Area Chair in several reputed conferences. He has completed international funded projects with France, Japan and Mexico. At the national level, he has been Principal Investigator of several consortium mode projects in the areas of machine translation, cross-lingual information access and treebank development. At present, he is Principal Investigator of an Indo-German SPARC project with the University of Saarlandes, Germany, on Multimodal Machine Translation and the Co-PI of several other international projects.

Prof. Valentina Emilia Balas is currently Full Professor in the Department of Automatics and Applied Software at the Faculty of Engineering, "Aurel Vlaicu" University of Arad, Romania. She holds a Ph.D. in Applied Electronics and Telecommunications from Polytechnic University of Timisoara. Dr. Balas is the author of more than 300 research papers in refereed journals and international conferences. Her research interests are in intelligent systems, fuzzy control, soft computing, smart sensors, information fusion, modeling and simulation. She is Editor-in-Chief to *International Journal of Advanced Intelligence Paradigms (IJAIP) and to International Journal of Computational Systems Engineering (IJCSysE),* an Editorial Board member of several national and international journals and is an evaluator expert for national, international projects and Ph.D. Thesis. Dr. Balas is Director of Intelligent Systems Research Centre in Aurel Vlaicu University of Arad and Director of the Department of International Relations, Programs and Projects in the same university. She served as General Chair of the International Workshop Soft Computing and Applications (SOFA) in eight editions 2005–2020 held in Romania and Hungary. Dr. Balas participated in many international conferences as Organizer, Honorary Chair, Session Chair and a member in Steering, Advisory or International Program Committees. She is a member of EUSFLAT, SIAM, a senior member of IEEE, a member in TC – Fuzzy Systems (IEEE CIS), Chair of the TF 14 in TC – Emergent Technologies (IEEE CIS), and a member in TC – Soft Computing (IEEE SMCS). Dr. Balas was past Vice-President (Awards) of IFSA International Fuzzy Systems Association Council (2013–2015) and is Joint Secretary of the Governing Council of Forum for Interdisciplinary Mathematics (FIM), A Multidisciplinary Academic Body, India.

Contributors

E. Baburaj Department of Computer Science Engineering, Marian Engineering College, Trivandrum, Kerala, India

Ashutosh Bajpai Indian Institute of Technology Hyderabad, Hyderabad, India

Balasubramanian Indian Institute of Technology Hyderabad, Hyderabad, India

Sivaji Bandyopadhyay Jadavpur University, Kolkata, India

Dhruba K. Bhattacharyya Tezpur University, Napaam, Tezpur, Assam, India

C. V. Bijitha Kannur, India

G. P. Biswas IIT(ISM), Dhanbad, Jharkhand, India

Amrita Chandra Jadavpur University, Kolkata, India

D. Christy Sujatha PG and Research Department of Computer Science, Rajah Serfoji Government College, Thanjavur, Tamil Nadu, India

Dipankar Das Jadavpur University, Kolkata, India

Debabrata Datta Homi Bhabha National Institute, Mumbai, India;
Bhabha Atomic Research Centre, Mumbai, India

C. G. Dethe UGC Academic Staff College, Nagpur, India

Jorge Diaz Universidad de la Costa (CUC), Barranquilla, Colombia

Ana María Echeverria Universidad de La Costa, Barranquilla, Colombia

J. Gnana Jayanthi PG and Research Department of Computer Science, Rajah Serfoji Government College, Thanjavur, Tamil Nadu, India

Jesús García Guliany Universidad Simón Bolívar, Barranquilla, Colombia

Pankaj Jadhav Homi Bhabha National Institute, Mumbai, India

Roberto Jimenez Universidad de la Costa, Barranquilla, Atlántico, Colombia

Jugal K. Kalita College of Engineering and Applied Science, University of Colorado, Colorado Springs, CO, USA

Simran R. Khiani SGBAU, Amravati, India;
GHRCEM, Pune, India

Aman Kumar LNMIIT, Jaipur, India

Ashwina Kumar DIT University, Dehradun, Uttarakhand, India

Naveen Kumar DIT University, Dehradun, Uttarakhand, India

Luis Luque Universidad de La Costa (CUC), Barranquilla, Colombia

Sainik Kumar Mahata Jadavpur University, Kolkata, India

T. Mathi Murugan Faculty of Computer Science and Engineering, Sathyabama Institute of Science and Technology, Chennai, India

Carlos Vargas Mercado Corporación Universitaria Latinoamericana, Barranquilla, Colombia

Prerna Mohit Indian Institute of Information Technology Senapati Manipur, Imphal, India

Siddhartha Mukhopadhyay Homi Bhabha National Institute, Mumbai, India;
Bhabha Atomic Research Centre, Mumbai, India

Naresh Babu Muppalaneni National Institute of Technology Silchar, Silchar, India

Harold Neira Universidad de La Costa (CUC), Barranquilla, Colombia

Luis Ortiz-Ospino Universidad Simón Bolívar, Barranquilla, Colombia

David Ovallos-Gazabon Universidad Simón Bolívar, Barranquilla, Colombia

Ripon Patgiri National Institute of Technology Silchar, Silchar, India

Shalini Ramanathan Department of Computer Science and Engineering, National Institute of Technology Tiruchirappalli, Tiruchirappalli, Tamil Nadu, India

Mohan Ramasundaram Department of Computer Science and Engineering, National Institute of Technology Tiruchirappalli, Tiruchirappalli, Tamil Nadu, India

Ivon Romero-Pérez Universidad Simón Bolívar, Barranquilla, Colombia

Carlos Rondón Universidad de la Costa (CUC), Barranquilla, Colombia

Manaswita Saikia Tezpur University, Napaam, Tezpur, Assam, India

Ernesto Steffens Sanabria Corporación Universitaria Latinoamericana, Barranquilla, Colombia

Joyeeta Singha LNMIIT, Jaipur, India

Anil Kumar Singh Assam Kaziranga University, Jorhat, Assam, India

Priyanka Singh Department of Computer Science and Engineering, SRM University-AP, Amaravati, Andhra Pradesh, India

Ranjeet Kumar Singh DIT University, Dehradun, Uttarakhand, India

K. Swarupa Rani School of Computer and Information Sciences, University of Hyderabad, Hyderabad, India

V. M. Thakare SGBAU, Amravati, India

Naushad Varish Department of Computer Science and Engineering Koneru Lakshmaiah Education Foundation, Guntur, Andhra Pradesh, India

Reynaldo Villareal–González Universidad Simón Bolívar, Barranquilla, Colombia

Luis Cabás Vásquez Corporación Universitaria Latinoamericana, Barranquilla, Colombia

Sai Kiran Wupadrasta Indian Institute of Technology Hyderabad, Hyderabad, India

Aizan Zafar School of Computer and Information Sciences, University of Hyderabad, Hyderabad, India

BicGenesis: A Method to Identify ESCC Biomarkers Using the Biclustering Approach

Manaswita Saikia, Dhruba K. Bhattacharyya, and Jugal K. Kalita

Abstract Biclustering has already been established as an effective tool to study gene expression data toward interesting biomarker findings for a given disease. This paper examines the effectiveness of some prominent biclustering algorithms in extracting biclusters of high biological significance toward the identification of interesting biomarkers. We have chosen Esophageal Squamous Cell Carcinoma (ESCC) as a case for our empirical study and our method called BicGenesis could identify eight genes as possible biomarkers for ESCC.

Keywords Biclustering · Biomarkers · ESCC · Regulatory network · Microarray · Hub-gene

1 Introduction

A clustering algorithm aims at finding sets of genes that exhibit similar variations on expression levels under all experimental conditions. That is to say, two different genes showing similar experimental tendencies across samples exhibit common patterns of regulation, reflecting interaction or the relation between their functions. The understanding of cellular processes however implicates that a subset of genes co-expressed under certain experimental conditions might possibly behave almost independently under other conditions. This resulted in the introduction of a two-mode clustering approach called biclustering or co-clustering [11] to group the genes and

M. Saikia (✉) · D. K. Bhattacharyya
Tezpur University, Napaam, Tezpur, Assam, India
e-mail: manaswitasaikia@gmail.com

D. K. Bhattacharyya
e-mail: dkb@tezu.ernet.in

J. K. Kalita
College of Engineering and Applied Science, University of Colorado, Colorado Springs, CO, USA
e-mail: jkalita@uccs.edu

R. Patgiri et al. (eds.), *Proceedings of International Conference on Big Data, Machine Learning and Applications*, Lecture Notes in Networks and Systems 180,
https://doi.org/10.1007/978-981-33-4788-5_1

conditions in both dimensions simultaneously. The use of clustering algorithms are often hindered by two limitations:

a. When computing gene similarity, most conditions or samples do not contribute information effectively resulting in an increase in the amount of background noise.
b. It has been observed that genes tend to participate in several functions which necessitates the need for assigning each gene to multiple clusters, which is not the case in exclusive clustering.

These limitations are overcome through the introduction of biclustering that finds subsets of genes exhibiting similar responses under a subset of conditions rather than all conditions. Genes might further exhibit varying regulation patterns under different contexts due to their participation in more than one function. Traditional clustering falls short in cases of cellular processes that are active only under specific conditions or in the case of genes that participate in differentially regulated multiple pathways. Biclustering techniques have the ability to cluster rows and columns of a data matrix simultaneously. Several biclustering algorithms [8–12, 15, 16, 28–30, 34–39] have been developed in recent years. This work performs a detailed experimental analysis of seven prominent biclustering techniques to analyze their effectiveness over gene expression data for ESCC patients. From our experimental study, we could identify eight interesting genes such as TP53, BRAF, HRAS, TUG1, MEG3, POLR2B, FXR1 and RHOA which have direct associations with ESCC.

2 Related Work

According to Pontes et al. [32], biclustering algorithms can be grouped as (i) algorithms based on evaluation measures that form biclusters according to the type of meta-heuristics used and (ii) non-metric biclustering that forms biclusters according to the most distinctive properties. The metric-based approach involves several biclustering strategies such as a iterative greedy [11, 21, 29, 37], stochastic greedy [8, 16, 36], nature inspired [10, 15, 26] and clustering based [23].

Non-metric biclustering algorithms do not use any evaluation measure as a guideline for the search. Graph-based approaches [35, 39] use the graph theory that either (a) uses nodes to represent elements of a bicluster, i.e., either genes or samples or both or (b) uses nodes to represent whole biclusters. Probabilistic algorithms [20, 24] use probabilistic models so as to extract biclusters from the dataset.

Table 1 Summary of the chosen biclustering algorithms

Approaches	Algorithms	Measure/technique	Dataset used
Iterative greedy search	Biclustering of expression data by Cheng and Church (CC) [11, 33]	Mean squared residue (MSR)	Human large B-cell lymphoma data [11]
Divide and Conquer	Bimax [34]	Binarization	Yeast saccharomyces cerevisiae cell cycle [5]
Graph-based approaches	QUalitative BIClustering algorithm (QUBIC) [25]	Discretization; bipartite graphs	E.coli [18], Arabidopsis [27], BRCA tumor [31]
Probabilistic models	Plaid models (PM) [24]	Sum of squared errors (SSE)	Yeast saccharomyces cerevisiae [5]
	Conserved gene expression Motifs (xMOTIFs) [30]	Conserved gene expression Motifs	Leukemia [19], Colon cancer [7], B-cell lymphoma [6]
Linear algebra-based approaches	Spectral biclustering (SB) [23]	Singular value decomposition (SVD)	Lymphoma [6], Leukemia [19]
	Iterative signature algorithm (ISA) [9]	SVD	Yeast saccharomyces cerevisiae [5]

3 Some Selected Biclustering Algorithms

In this section, we present seven biclustering algorithms chosen based on their behavior, popularity and effectiveness in handling gene expression data. The following is a brief description of each algorithm (Table 1).

(a) *Cheng and Church (CC)* [11] biclustering was the first effort to apply biclustering to gene expression data. This work [11] attempts to find n biclusters in an expression data matrix using sequential covering strategy. The most significant contribution of this work is the formulation of Mean Squared Residue (MSR) [11], a measure to assess the quality of a bicluster of expression data. As the name suggests, MSR uses the means of genes and condition expression values to evaluate the coherence of the genes and conditions in a bicluster α [11].

$$MSR(\alpha) = \frac{1}{|I| \cdot |J|} \sum_{i=1}^{i-|I|} \sum_{j=1}^{j-|J|} (a_{ij} - a_{iJ} - a_{Ij} + a_{IJ})^2 \quad (1)$$

where a_{ij}, a_{iJ}, a_{Ij} and a_{IJ} represent the elements in the ith row (condition) and jth column (gene), the row and column means and the mean of α, respectively [11]. MSR is widely considered to be the earliest quality metric to be defined for biclusters of expression data [32]. However, because MSR is unable to capture shifting tendencies, many approaches have incorporated modified versions of MSR.

(b) In *Iterative Signature Algorithm (ISA)* [9], biclusters are primarily defined as transcription modules retrievable from the expression data [32, 33]. A set of co-regulated genes alongside a set of conditions where co-regulation is the most stringent [32] constitutes a transcription module (TM). TMs in the data are found by applying a generalized Singular Value Decomposition (SVD). In a module, the gene and condition similarity is determined by two thresholds which in turn regulate its size. The algorithm starts with a random selection of a set of genes or conditions which are then iteratively refined until they match the definition of a TM producing one bicluster for every iteration. As the initial selection of seeds are random and lack overlap restrictions, the resulting biclusters might have overlapped genes and/or conditions.

(c) The *QUalitative BIClustering algorithm (QUBIC)* [25] initially represents the input data in the form of a matrix with integer values either qualitatively or semi-qualitatively. Under a subset of conditions, if two rows of the matrix have identical integer values then the two corresponding genes can be considered as correlated. The qualitative (or semi-qualitative) representation helps the algorithm in detecting different kinds of biclusters such as scaling patterns. Furthermore, it is well suited for the detection of positive and negative correlation expression patterns. The algorithm initiates by taking into account the qualitative or semi-qualitative matrix with vertices for genes and constructing a weighted graph. Edges that connect every pair of genes however have a corresponding weight computed based on the similarity between the two corresponding rows. Initially, a bicluster is built using the heaviest unused edge as the seed, and the algorithm proceeds to add additional genes to the current solution iteratively.

(d) *Bimax* [34] enumerates all the possible biclusters in the input data matrix by employing a divide and conquer strategy. As the algorithms tend to seek rectangles of 1's in a binary matrix, input data has to be binarized. The algorithm begins by choosing any row with a mixture of 0's and 1's. The input data matrix might either fit the criteria or it might be the case that no such row exists. This is the case under two circumstances: (a) all entries in the matrix are 1, i.e., the entire matrix is a single bicluster or (b) all entries in the matrix are 0, i.e., no bicluster exists. The algorithm starts by choosing the first row, $r*$, arbitrarily from the input $m \times n$ matrix; $M \cdot r*$ is then used to subdivide M into two submatrices each of which are processed independently. Submatrices are found by dividing the columns $C = 1, \ldots, n$ into two sets: $C_U = C : M[r^*, c] = 1$ and $C_V = C - C_U$ [3]. The algorithm then proceeds to divide the m rows into three sets: (1) R_U: Rows with 1s only in C_U, (2) R_W: Rows with 1s in both C_U and C_V and (3) R_v: Rows with 1s only in C_V [3]. The rows and columns of M are rearranged to make each set contiguous thus resulting in the following observations: (a) Submatrix by (R_U, C_V) cannot form any bicluster due to the fact that they are empty and (b) Submatrix $U = (R_U \cup R_W, C_U)$ and submatrix $V = (R_W \cup R_V,$

$C_U \cup C_V$) contain all possible biclusters in M [3]. This procedure continues by recursively processing U and then V until no rows with mixed 0's and 1's exist.

(e) In *xMOTIFs (conserved gene expression Motifs)* [30], a subset of genes is an xMOTIF if it satisfies the following conditions: (a) the subset is simultaneously conserved across a subset of samples, and (b) in the subset, across a set of samples, the gene expression level is conserved if it is in the same state in each of the samples [32]. For every gene, the fundamental requirement is a record of intervals corresponding to the states in which the corresponding gene is expressed in the samples [32, 33]. However, in order to avoid finding very small or very large xMOTIFs, some constraints are placed on the conservation, maximality and size over its formal definition. This work [30] aims toward computing the largest xMOTIF using a probabilistic algorithm that exploits the mathematical structure of xMOTIF [32]. With the aim of finding several xMOTIFs in the data, samples that meet the expectation of each xMOTIF are iteratively removed followed by a search for the new largest xMOTIF until all samples adhere to some xMOTIF.

(f) *Plaid models* [24] represent the gene-condition matrix as a superposition of layers, corresponding to biclusters [32, 33]. In this work, Lazzeroni and Owen allow a gene to be in more than one bicluster and have described several versions of the model. The most general model is defined as [24, 32]

$$X_{ij} = \sum_{k=0}^{K} (\theta_{ijk} - \sigma_{ijk} - \kappa ijk) \tag{2}$$

where X_{ij} = the expression level of ith gene in jth sample, κ = number of biclusters, θ_{ij0} = the background layer and θ_{ijk} = four types of models, corresponding to the type of biclusters (overlapped, exclusive, ...). Each $\sigma_{ik} \in 0, 1 = \{1, \text{if } i\text{th}$ gene is in the kth bicluster; 0 otherwise$\}$ [32]. Each $\kappa - jk \in 0, 1 = \{1, \text{if } j\text{th}$ sample is in the kth bicluster; 0 otherwise$\}$. This greedy algorithm finds k biclusters through the addition of one layer at a time by seeking a plaid model that minimizes the sum of squared errors during the process of approximating the data matrix to the model.

(g) *Spectral Biclustering (SB)* [23] was designed specifically for analyzing microarray cancer datasets. The basic assumption is that with blocks of high expression as well as low expression levels, the expression matrix conforms to a checkerboard-like structure. Hence, this approach primarily focuses on finding these distinct patterns using Singular Value Decomposition (SVD) and eigenvectors. The search further incorporates the normalization of genes. SB prevents overlapping among biclusters and also makes attempts in ensuring that every gene and condition are included in at least one bicluster.

Initially, our method called BicGenesis (Bicluster enabled Gene Expression Analysis) considers a large number of synthetic datasets generated based on [17] to evaluate the performance of the seven chosen biclustering algorithms. A few algorithms, namely Bimax, Plaid, xMotif, Cheng and Church, and Spectral tend to exhibit excessive running time in case of carelessly chosen parameters. However, none of these

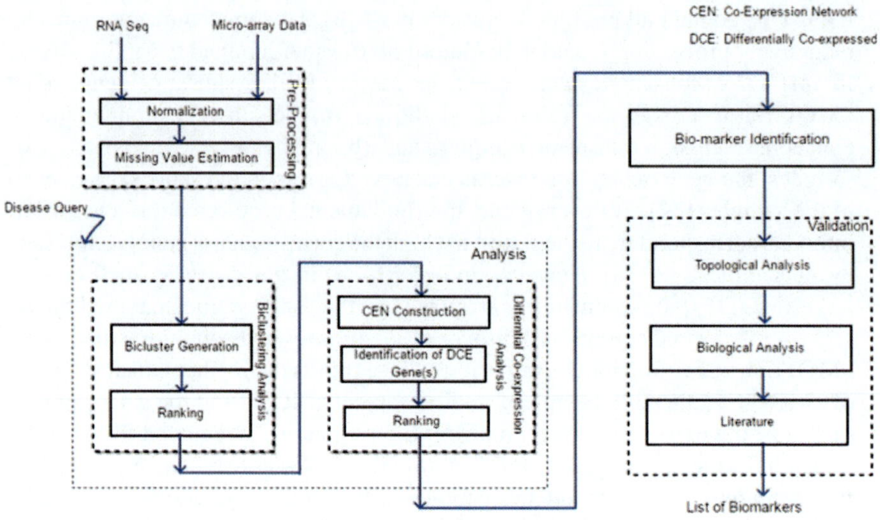

Fig. 1 Proposed framework

algorithms were able to fully separate biclusters with substantial overlap. Most of the algorithms, with the exception of *Bimax*, which performed significantly well in all scenarios, are found to be affected by noise. It is important to note that each algorithm performs well on different biclustering models.

4 Proposed Framework

We have proposed a framework (Fig. 1) for detection of biomarkers using biclustering approach. Irrespective of whether the data is RNAseq or microarray, there is a need for preprocessing. It is the aim to explore the best practices to preprocess gene expression data to help the extraction of interesting patterns from the gene expression data. One important preprocessing task is the normalization of the input data. Similarly, one might also encounter missing values in the input data. The preprocessing unit is also responsible for missing value estimation. The analysis unit works toward identifying the particular gene(s) as the cause of the disease. However, we have to consider that genes tend to exhibit varying behaviors. Keeping this in mind, we plan on designing a method that is based on different analysis approaches. Several efforts have been made to develop a competent algorithm for Biclustering Analysis. It is the aim to explore the possibility of (i) enhancing an existing biclustering technique and (ii) developing a new biclustering technique to enable identifying biologically significant factors using biclustering to support disease diagnosis.

The first step toward bicluster analysis is the application of an enhanced biclustering algorithm on the preprocessed data which results in the generation of biclusters. These biclusters are then ranked based on some measures that are experiment specific. Based on the score, relevant bicluster(s) is/are chosen for further analysis. Differential co-expression analysis is an improvement on the conventional method of analyzing differentially expressed genes due to its ability to discover disease mechanisms and underlying regulatory dynamics [13]. Naturally, the first step toward differential co-expression analysis is the construction of Co-Expression Network (CEN). Many algorithms have been proposed to identify differentially co-expressed (DCE) gene sets and quantify differential co-expressions. We have specifically selected Weighted Graph Co-expression Network Analysis (WGCNA) [38]. These genes are then ranked based on experiment-specific measures and the genes with the highest scores are chosen. Based on the acquired knowledge of certain gene module(s)/gene(s) as a significant cause in the context of the disease as established by the analysis unit, the biomarker identification unit works on identifying interesting biomarker(s). The validity of the identified biomarkers are assessed based on the following aspects:

a. Topological analysis of the identified modules might lead to the discovery of causal genes which are otherwise not associated with that particular disease. There might be a scenario where two genes say, g_x and g_y, in a well-connected graph have an edge between them, and g_x might be already known to be a primary or elite gene. In that case, there is a high probability that g_y is a causal gene.
b. By constructing a biological network (for example, Gene Regulatory Network) on potential biomarkers identified, one can analyze the association patterns (such as regulatory behavior) among the genes and the corresponding Transcription Factors (TFs).
c. Established literature with respect to the given disease can help to validate the identified biomarkers by checking their direct or indirect association with the disease.

5 Performance Evaluation of BicGenesis

5.1 Benchmark Dataset

From ESCC [14], GSE20347 has been chosen as the gene expression dataset for experimental study. To characterize the dataset in esophageal squamous cell carcinoma, a gene expression was examined in a tumor and it matched normal adjacent tissues from 17 ESCC patients from a high-risk region of China [14]. The specification of the dataset is available at [14].

5.2 Results on GSE20347

The aforementioned 7 chosen algorithms were experimented on biological data, specifically on the ESCC dataset. The evaluation of the behavior of the algorithms on only synthetically generated data may fail to highlight many aspects of the chosen algorithms. The unpredictable behavior of biological data may expose various aspects of an algorithm. In spite of this concern, we proceed with our approach with the idea that a few of these algorithms might be able to successfully detect causal genes for ESCC as listed in MalaCard [4]. The algorithms were run with carefully (heuristically) chosen parameters. A brief summary of the experimental results is presented in Table 2. The Cheng and Church algorithm finds a single bicluster for the GSE20347 dataset with all genes and conditions (22,278 genes and 34 conditions) rendering the bicluster useless due to the fact that all genes are present regardless. Similarly, Spectral finds one bicluster (10 genes and 10 conditions) with only one of the five causal genes which may be due to excessive pruning leading to loss of significant information. For Plaid and QUBIC, the detected biclusters are small thus making it easier for further analysis. However, as in the case of Spectral, due to pruning none of the causal genes present in the dataset were detected. FABIA found 13 biclusters, 5 of which were deemed significant due to the presence of two causal genes HRAS and TP53. Bimax found 10 medium-sized biclusters, with approxi-

Table 2 Summary of implementation on ESCC dataset

Algorithm	No. of biclusters	Causal genes detected	Cluster no.
Bimax [34]	10	TP53	7, 6, 10
		TUG1	1, 4, 5, 6, 7, 10
		HRAS	4, 5, 6, 7, 8, 10
		MEG3	2, 3
FABIA [22]	13	HRAS	3
		TP53	3.4.9.10.11
QUBIC	100	None	NA
ISA [9]	27	BRAF	4, 18, 20
		TP53	12, 15, 19, 21, 22, 24, 26, 27
		HRAS	18, 20
		TUG1	15, 21, 27
xMotif [30]	10	BRAF	1
		HRAS	1
		TP53	1, 7
		TUG1	1, 2
		MEG3	1
Plaid [24]	3	None	NA
Spectral [23]	1	MEG3	1

mately 5500 genes and 2 conditions, 9 of which detected 4 causal genes. ISA found 27 relatively smaller sized biclusters making it useful for further analysis. Four of the causal genes were detected in these biclusters. xMotif was successful in detecting 5 out of 6 causal genes present in the dataset but the bicluster sizes varied significantly for each bicluster. It is to be noted that most of these causal genes were present in the first cluster, which is the largest bicluster with 18,824 genes.

5.3 *Topological Study*

From Table 2, it can be concluded that 29 biclusters can be considered relevant for further analysis. Taking into consideration that some of the obtained biclusters are significantly large (e.g., xMotif bicluster 1 consists of 17,864 genes) thus makes them difficult for further analysis. Moreover, it is important to get an idea about how the genes interact with each other so as to get an in-depth idea about the dynamics within the bicluster. This is achieved through Weighted Graph Co-Expression Network Analysis (WGCNA) [38]. The Weighted Graph Co-expression Network is represented as an adjacency matrix with assigned weight for all edges. We assume that weights above a threshold value (in this case, threshold $= 0.5$) signifies the existence of an edge. In other words, two genes are considered to be co-expressed if their correlation value is ≥ 0.5. We represent all such edges within each bicluster in the form of an upper triangular matrix with value 1 if an edge exists and 0, otherwise.

5.4 *Biological Study*

It is an acknowledged fact that some genes play more significant roles than others within a bicluster. We consider hub-genes, i.e., genes with relatively high connectivity, as genes with relevant information and turn our focus on them. For each of the relevant biclusters, we extract 10 such hub-genes. After successfully finding a list of 150 hub genes from mapped biclusters shown, we construct a Regulatory Network (RN) with the subset of hub-genes and connected TFs. 23 of the hub-genes are Transcription Factors (TFs). The corresponding RN is obtained as an adjacency list with directed edges from TFs to other hub-genes with a weight assigned for each existing edge. However, out of the 6463 such extracted edges, a significant number had negligible weight. As such we filter out such edges using a threshold weight of 0.1 to get an RN with 501 edges. Previously, we extracted relevant biclusters based on the causal genes listed in MalaCard [4] for ESCC. We then further proceed with the analysis of all genes found relevant in the previous step. As mentioned earlier, 5 causal genes, namely TP53, TUG1, MEG3, HRAS and BRAF, for ESCC as listed in MalaCard [4], have been detected in the biclusters. Furthermore, intOgen [2] estab-

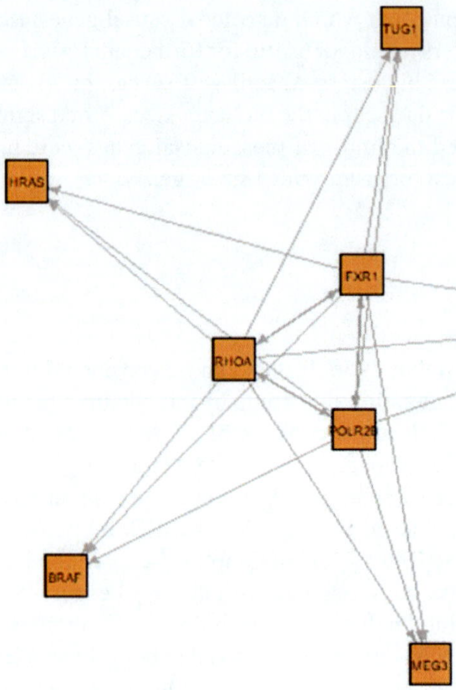

Fig. 2 Gene regulatory network based on the potential biomarker genes for ESCC

lishes that 3 of these causal genes, namely TP53, BRAF and HRAS, are known drivers for 25, 7 and 2 types of cancer, respectively. While MEG3 and HRAS are not known drivers according to IntoGene, a few works say otherwise. CBioportal also lists a few of the TFs, namely MATR3, TAX1BP1, FXR1, TTLL4 and TP53BP1, as possible causal genes. It also is to be noted that CBioPortal solidifies the claim that BRAF and TP53 are causal genes. Furthermore, according to intOgen [2] mutations in 2 TF genes, FXR1 and POLR2B possibly contribute to 6 and 7 cancer types, respectively. According to intOgen [2], another TF RHOA possibly contributes to cancer-type Stomach adenocarcinoma. Figure 2 is the RN for all 8 genes. We observe that all 3 TFs, namely FXR1, POLR2B and RHOA, play a role in regulating other non-TF genes, i.e., TP53, BRAF, HRAS, TUG1 and MEG3. Furthermore, all the TFs regulate one another which is represented by a double directed edge in Fig. 2. Table 3 gives a summary of a few of the prominent hub-genes.

Table 3 Summary of a few prominent hub-genes

Gene name	Highest degree	TF?	Known driver?/Causal gene in ESCC	Edge to causal gene
FXR1	1332	Yes	Possibly/Possibly	TP53
TP53BP1	1188	Yes	Possibly/Possibly	TP53
MATR3	1470	Yes	No/No	TP53, POLR2B
BZW1	147	Yes	No/Possibly	None
TAX1BP1	705	Yes	No/Possibly	None
TTLL4	536	Yes	No/Possibly	TP53
POLR2B	1326	Yes	Possibly/No	TP53
RHOA	76	Yes	Possibly/Possibly	RHOA
RPL10A	1492	No	No/No	TP53, POLR2B
CTSK	181	No	No/Possibly	TP53
RPL31	261	No	No/Possibly	FXR1
RPL11	450	No	No/No	TP53, POLR2B, FXR1
CNN3	1147	No	No/No	None
LAMC1	788	No	No/No	TP53, POLR2B
ENO1	712	yes	No/No	TP53, POLR2B
RPS5	1490	No	No/No	TP53, POLR2B
GAS6	1488	No	No/No	TP53, POLR2B
RPL30	1487	No	No/No	TP53, POLR2B
BCL2	160	Yes	No/No	TP53

5.5 *Some Interesting Biomarkers Identified by BicGenesis*

We introduce each of these eight genes in Table 4.

6 Conclusion

All the chosen algorithms are very costly with regards to running time due to the fact that an exhaustive search is performed to identify significant biclusters. However, there is scope to overcome this using appropriate implementation. Hub-gene centric analysis of the obtained relevant biclusters helps in identifying interesting biomarkers from the constructed biological networks. We observe that 8 genes related to cancer are found to be highly significant in ESCC. Extensive validation is going on with these nine potential biomarkers to establish their critical role in ESCC.

Table 4 Summary of the detected genes

Gene name	Characteristic	Diseases
TP53 [1]	Encodes a tumor suppressor protein containing transcriptional activation, DNA binding and oligomerization domains [1]	Esophageal carcinoma, Lung adenocarcinoma, **Lung squamous cell carcinoma**, **Head and neck squamous cell carcinoma**, Thyroid carcinoma and 20 more [2]
BRAF [1]	Encodes a protein belonging to the RAF family of serine/threonine protein kinases; regulating the MAP kinase/ERK signaling pathway, which affects cell division, differentiation and secretion [1]	Prostate adenocarcinoma, **Lung adenocarcinoma**, Thyroid carcinoma, Cutaneous melanoma, Multiple myeloma, Colorectal adenocarcinoma and Glioblastoma multiforme [2]
HRAS [1]	Belongs to the Ras oncogene family, whose members are related to the transforming genes of mammalian sarcoma retroviruses; products encoded by these genes function in signal transduction pathways [1]	Follicular thyroid cancer, bladder cancer, **Oral squamous cell carcinoma**, Thyroid carcinoma and **Head and neck squamous cell carcinoma** [2]
TUG1 [1]	Produces a long non-coding RNA which promotes cell proliferation and is upregulated in tumor cells [1]	Intrahepatic Cholangiocarcinoma, Hepatoblastoma, **Squamous cell carcinoma**, osteogenic sarcoma and bladder urothelial carcinoma [2]
MEG3 [1]	Multiple alternatively spliced transcript variants, long non-coding RNAs (lncRNAs), are transcribed from MEG3 [1]	Pituitary carcinoma, testicular germ cell cancer, pancreatic endocrine carcinoma, **Vulva squamous cell carcinoma** and gastric cardia adenocarcinoma [2]
POLR2B [1]	Encodes the second largest subunit of RNA polymerase II (Pol II), a DNA-dependent RNA polymerase, snRNA and microRNA [1]	**Lung adenocarcinoma**, Breast carcinoma, Cutaneous melanoma, Serous ovarian adenocarcinoma, Stomach adenocarcinoma, Colorectal adenocarcinoma and **Head and neck squamous cell carcinoma** [2]
FXR1 [1]	Encodes an RNA binding protein that interacts with the functionally-similar proteins FMR1 and FXR2 [1]	Lung adenocarcinoma, **Lung squamous cell carcinoma**, Colorectal adenocarcinoma, Cutaneous melanoma, Uterine corpus endometrioid carcinoma and Multiple myeloma [2]
RHOA [1]	Encodes a member of the Rho family of small GTPases; Overexpression of RHOA has been associated with tumor cell proliferation and metastasis [1]	Colorectal cancer, Stomach adenocarcinoma, **Squamous cell carcinoma** and Peripheral t-cell lymphoma [2]

References

1. Genecards. https://www.genecards.org/, 31 May 2019
2. Intogen. https://www.intogen.org/, 31 May 2019
3. Kemaleren. http://www.kemaleren.com/post/bimax/, 31 May 2019
4. Malacards. https://malacards.org/, 9 Dec 2018
5. Yeast saccharomyces cerevisiae cell cycle expression dataset. http://arep.med.harvard.edu/biclustering
6. Alizadeh AA, Eisen MB, Davis RE, Ma C, Lossos IS, Rosenwald A, Boldrick JC, Sabet H, Tran T, Yu X et al (2000) Distinct types of diffuse large b-cell lymphoma identified by gene expression profiling. Nature 403(6769):503
7. Alon U, Barkai N, Notterman DA, Gish K, Ybarra S, Mack D, Levine AJ (1999) Broad patterns of gene expression revealed by clustering analysis of tumor and normal colon tissues probed by oligonucleotide arrays. Proc Natl Acad Sci 96(12):6745–6750
8. Angiulli F, Cesario E, Pizzuti C (2008) Random walk biclustering for microarray data. Inf Sci 178(6):1479–1497
9. Bergmann S, Ihmels J, Barkai N (2003) Iterative signature algorithm for the analysis of large-scale gene expression data. Phys Rev E 67(3):031902
10. Bryan K, Cunningham P, Bolshakova N (2006) Application of simulated annealing to the biclustering of gene expression data. IEEE Trans Inf Technol Biomed 10(3):519–525
11. Cheng Y, Church GM (2000) Biclustering of expression data. In: ISMB, vol 8, pp 93–103
12. Chi EC, Allen GI, Baraniuk RG (2017) Convex biclustering. Biometrics 73(1):10–19
13. Chowdhury HA, Bhattacharyya DK, Kalita JK (2019) (Differential) co-expression analysis of gene expression: a survey of best practices. IEEE/ACM Trans Comput Biol Bioinform 17(4):1154–1173
14. Clifford RJ, Hu N, Lee MP, Taylor PR (2011) Analysis of gene expression in esophageal squamous cell carcinoma ESCC. NCBI
15. Coelho GP, de França FO, Von Zuben FJ (2009) Multi-objective biclustering: when non-dominated solutions are not enough. J Math Model Algorithm 8(2):175–202
16. Dharan S, Nair AS (2009) Biclustering of gene expression data using reactive greedy randomized adaptive search procedure. BMC Bioinform 10(1):S27
17. Eren K, Deveci M, Küçüktunç O, Çatalyürek ÜV (2012) A comparative analysis of biclustering algorithms for gene expression data. Brief Bioinform 14(3):279–292
18. Faith JJ, Driscoll ME, Fusaro VA, Cosgrove EJ, Hayete B, Juhn FS, Schneider SJ, Gardner TS (2007) Many microbe microarrays database: uniformly normalized Affymetrix compendia with structured experimental metadata. Nucl Acids Res 36(Suppl 1):D866–D870
19. Golub TR, Slonim DK, Tamayo P, Huard C, Gaasenbeek M, Mesirov JP, Coller H, Loh ML, Downing JR, Caligiuri MA et al (1999) Molecular classification of cancer: class discovery and class prediction by gene expression monitoring. Science 286(5439):531–537
20. Gu J, Liu JS (2008) Bayesian biclustering of gene expression data. BMC Genomics 9(1):S4
21. Hartigan JA (1972) Direct clustering of a data matrix. J Am Stat Assoc 67(337):123–129
22. Hochreiter S, Bodenhofer U, Heusel M, Mayr A, Mitterecker A, Kasim A, Khamiakova T, Van Sanden S, Lin D, Talloen W et al (2010) Fabia: factor analysis for bicluster acquisition. Bioinformatics 26(12):1520–1527
23. Kluger Y, Basri R, Chang JT, Gerstein M (2003) Spectral biclustering of microarray data: coclustering genes and conditions. Genome Res 13(4):703–716
24. Lazzeroni L, Owen A (2002) Plaid models for gene expression data. Stat Sinica 61–86
25. Li G, Ma Q, Tang H, Paterson AH, Xu Y (2009) Qubic: a qualitative biclustering algorithm for analyses of gene expression data. Nucleic Acids Res 37(15):e101–e101
26. Liu J, Li Z, Hu X, Chen Y (2009) Biclustering of microarray data with mospo based on crowding distance. BMC Bioinform10:S9; BioMed Central (2009)
27. Lowry DB, Logan TL, Santuari L, Hardtke CS, Richards JH, DeRose-Wilson LJ, McKay JK, Sen S, Juenger TE (2013) Expression quantitative trait locus mapping across water availability environments reveals contrasting associations with genomic features in arabidopsis. Plant Cell 25(9):3266–3279

28. Mandal K, Sarmah R, Bhattacharyya DK (2018) Biomarker identification for cancer disease using biclustering approach: an empirical study. IEEE/ACM Trans Comput Biol Bioinform 16(2):490–509
29. Mukhopadhyay A, Maulik U, Bandyopadhyay S (2009) A novel coherence measure for discovering scaling biclusters from gene expression data. J Bioinform Comput Biol 7(05):853–868
30. Murali T, Kasif S (2002) Extracting conserved gene expression motifs from gene expression data. In: Biocomputing 2003. World Scientific, pp 77–88
31. Network CGA et al (2012) Comprehensive molecular portraits of human breast tumours. Nature 490(7418):61
32. Pontes B, Giráldez R, Aguilar-Ruiz JS (2015) Biclustering on expression data: a review. J Biomed Inform 57:163–180
33. Pontes Balanza B (2013) Evolutionary biclustering of gene expression data shifting and scaling pattern-based evaluation
34. Prelić A, Bleuler S, Zimmermann P, Wille A, Bühlmann P, Gruissem W, Hennig L, Thiele L, Zitzler E (2006) A systematic comparison and evaluation of biclustering methods for gene expression data. Bioinformatics 22(9):1122–1129
35. Tanay A, Sharan R, Shamir R (2002) Discovering statistically significant biclusters in gene expression data. Bioinformatics 18(Suppl 1):S136–S144
36. Yang J, Wang H, Wang W, Yu PS (2005) An improved biclustering method for analyzing gene expression profiles. Int J Artif Intell Tools 14(05):771–789
37. Yip KY, Cheung DW, Ng MK (2004) HARP: a practical projected clustering algorithm. IEEE Trans Knowl Data Eng 16(11):1387–1397
38. Zhang B, Horvath S (2005) A general framework for weighted gene co-expression network analysis. Stat Appl Genet Mol Biol 4(1)
39. Zhao L, Zaki MJ (2005) Microcluster: efficient deterministic biclustering of microarray data. IEEE Intell Syst 20(6):40–49

An Improved Energy-Aware Secure Clustering Technique for Wireless Sensor Network

Simran R. Khiani, C. G. Dethe, and V. M. Thakare

Abstract With the recent advancements in technology, nowadays wireless sensor networks are the most prominent networks deployed almost in all the areas such as healthcare, agriculture, defense, etc. Wireless sensor nodes are basically used to monitor and send the data to a central machine for analysis purposes. The major concern while deploying the wireless sensor network is the limited computational resources and battery-operated tiny sensor nodes. Hence, the organization of the network must take into account these constraints. This chapter proposes a novel clustering algorithm which works on different variables that select the most prominent node as a Cluster Head. Due to the changes in the underlying network, Type 2 fuzzy logic system is used in the clustering to generate the rule base to develop a feasible system. The size of the cluster is unequal as according to the survey, the unequal distribution of nodes consumes less energy as compared to equal distribution. The results of the simulation show that the proposed system prolongs the lifespan of the network.

Keywords Dynamic clustering · Wireless sensor network · Fuzzy logic · Cluster head · Aggregation

1 Introduction

Wireless Sensor networks have a large number of tiny sensor nodes scattered in a huge area. These nodes can sense parameters such as soil moisture, temperature, pressure, etc. There are many applications such as monitoring of environment, defense work,

S. R. Khiani (✉) · V. M. Thakare
SGBAU, Amravati, India
e-mail: rashmi.makhijani2002@gmail.com

S. R. Khiani
GHRCEM, Pune, India

C. G. Dethe
UGC Academic Staff College, Nagpur, India

R. Patgiri et al. (eds.), *Proceedings of International Conference on Big Data, Machine Learning and Applications*, Lecture Notes in Networks and Systems 180,
https://doi.org/10.1007/978-981-33-4788-5_2

tracking of objects, etc. But, since the wireless sensor network has limited power, less memory, and little computation capability, it is difficult to develop applications based on WSNs.

The major difficulty in designing WSN network is the energy of the battery. The battery of sensor nodes cannot be changed after the deployment, so battery power should be less consumed in order to enhance the lifetime of the network. There are many algorithms developed in order to reduce the energy consumed by the network. Clustering [1] technique is one of the most commonly used methods for this purpose. In the clustering method, the network is divided into various groups based on certain attributes. In the first phase, the network is formed and depending upon the energy of the nodes, the node with the highest energy is selected as the Cluster Head (CH). After the CH is selected, the member nodes nearby the CH will join the network. The member nodes will sense the environment parameters and transmit the data to the CH and will go to an idle state. The responsibility of the CH is to aggregate all the data and send it to the Base Station (BS). The advantage of this method is that the CH is only transmitting the data hence energy will be less consumed.

1.1 Routing Protocols

Routing is also an important criterion in transmitting data from one node to the other node. It is basically used to select the best available path from source to destination. Each node will initially send its information such as node id, energy, and distance to the BS. This concept works similarly as in computer network. The routing data available at the BS will be used to select the best available path out of many paths. The selected path will be used to send data to the BS. Routing protocols in WSN are divided based on the operation of the protocol and structure of network.

In flat routing, all the sensor nodes have the same energy and sensing capability, whereas in hierarchical routing, nodes are divided based on the energy level and each node is assigned a different task. The low-level nodes perform sensing and the nodes at higher level perform the task of collecting and transmitting the data to the BS (Fig. 1).

1.2 Clustering

As the main challenge in WSN is battery power of the sensor nodes, many techniques are implemented to improve the energy consumption of the network. Clustering is a very important technique in increasing the life of the network. Clustering means dividing the network into groups and electing a node as a CH. The clustering algorithm can be classified into two types: Centralized and Distributed. In the centralized method, a central BS runs the algorithm to elect CH, whereas in the distributed

Fig. 1 Routing protocols

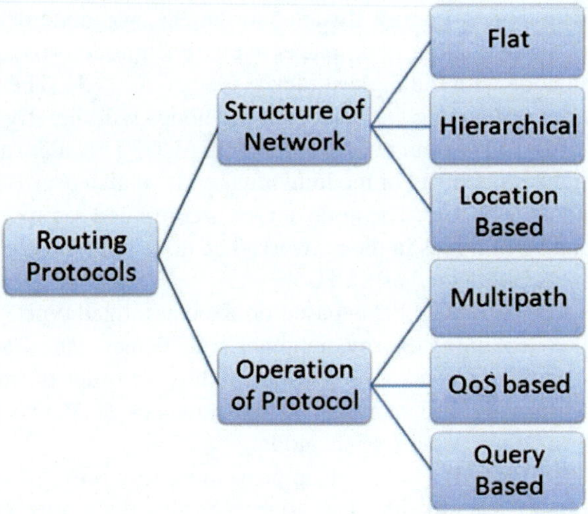

method, there is no central authority to select CH, the individual sensor nodes elect the CH based on accessibility.

The first clustering algorithm developed was named LEACH [2] in 2000. It uses a distributed method and depends upon the possibility of a node to become CH. When a node selects itself as CH, it sends a joining message to all other nodes. The nodes nearby join the group. The CH will be selected for some predefined amount of time. As this algorithm is simple to implement and less costly, it is the most famous and relevant algorithm used even in today's study work.

LEACH algorithm is further implemented with some important parameters [3] such as remaining energy in order to easily select CH. R-LEACH [4] is another version in which a random number is generated and compared with a threshold in order to select a CH. Similarly, some other methods [5] are also used to create random number which may be used to select a CH. Hybrid method is also used which checks the chance of each node or individual node to be selected as CH in distributed [6, 7] as well as centralized [8, 9] environment. In the centralized method, additional hardware is required for getting the dynamic changes in the connectivity of the network. This incurs extra energy consumption by each sensor node as data is sent to the BS for a central decision. So more complex algorithms need to be designed in order to lessen the exchange of data. In order to avoid the errors in selection process, fuzzy logic Type 1 and Type 2 are introduced by Gupta which was the first clustering algorithm based on fuzzy logic in WSNs. This algorithm is a centralized algorithm that uses three parameters, namely remaining energy, neighbor nodes, and centrality as input to an expert system to find out the CH. But this method needs the exact location of the sensor node and the details of the remaining energy of the node continuously.

The improvement to this algorithm was type 1-based FUCA [6] and MOFCA [10] which are distributed algorithms. Each node has a tentative probability to be selected as CH having clusters of different sizes. This method uses parameters such as remaining energy, distance to the BS, and node density as input and the output are a range and a rank, respectively. Each node broadcasts its rank in its range and the node with the highest rank is elected as a CH. There may be a problem with the nodes not having any rank so these nodes will directly send the data to the BS.

FBUC [7] is an improvement to EAUCF [11] algorithm and selects the CH based on the probability of the individual node. It also uses BS distance, remaining energy, and node degree. The node degree is computed as no of desired nodes divided by the total no of nodes in the network. The node with the highest remaining energy within a range is elected as a CH.

CRT2FLACO [9] is based on the centralized type 2 Mamdani fuzzy logic which takes remaining energy, neighbor node density, and BS distance as input and output is probability and a competition radius. In order to transmit the data is BS, all the CH form a chain using ACO. The drawback of ACO is that it needs location in order to calculate density of the node.

CHEETAH [12] is an improvement to CRT2FLACO algorithm. This is a type 2 fuzzy system that uses four variables remaining energy, BS distance, average number of nodes in the time period when the node was elected as CH, and efficiency. Efficiency is calculated as power saved by all the member nodes of a cluster after sending data to their CH. Another improvement [13] in the method is implemented by applying scheduling method in order to increase the network time.

2 Proposed System

In order to improve the energy consumption of the network, the proposed work is divided into three phases, namely network formation, CH selection, and data transmission.

Phase I: In the first phase, all the sensor nodes will be deployed randomly in the field area and connections between the nodes are formed. All the nodes in the network are homogenous, i.e., all have same capability and battery power. The location of the BS and all other nodes is stationary, i.e., once deployed the location of the nodes will not change.

After the network is formed, the BS will send control signal to all the nodes. As a reply to the control signal from BS, all the nodes will send their information such as energy, id, and distance to the BS which will be used in deciding the CH.

Phase II: The BS applies fuzzy logic in the second phase to elect the CH and creates a CH list. Fuzzy logic is used in the proposed method. Centralized algorithm is used by the BS for selecting CH as the BS has a large memory space and more powerful than the sensor nodes.

As many parameters [14] such as energy of node, distance to the BS, neighboring nodes, and intra-cluster distance, latency, delay are available for selecting the CH. It is a difficult task to select parameters for choosing a CH.

The proposed work uses remaining energy of the node, distance to the BS, neighboring nodes, and intra-cluster distance for choosing CH. Energy of a node [21] is a very important parameter. During data transmission and reception, energy of a node is consumed and range of the energy is from 1 to 3 J. As all the data is to be transmitted to the BS, the position of the BS is also important. The distance between the BS and a node is calculated using Euclidean distance. Distance and energy consumption during data transmission are directly proportional to each other i.e. if the distance is less, less energy will be consumed.

$$\text{Dist\_Euclidean} = \sqrt{(x_i - x_c)^2 + (y_i - y_c)^2},$$

where x_i and y_i are the x and y coordinates of a sensor node, respectively, and x_c and y_c are the coordinates of the BSs.

Similarly, the intra-cluster distance is also used and it is calculated by taking the average of distance of all the nodes.

$$\text{Dist\_IntraCluster} = \frac{1}{\mathrm{N}} \sum_{1}^{\text{no\_c}} \sum_{1}^{\text{no\_s}} \|(x - c_i)\|^2,$$

where no_c = no. of clusters, no_s = no. of sensors in a cluster, x = sensor node, c_i = center of a cluster.

After calculating all the parameter values, Type-2 Mamdani FLS (T2MFLS) [15–17] technique is used for electing a CH. Fuzzy system has various components such as inference engine, inference rules, fuzzy logic, and de-fuzzification. The input parameters such as energy, intra-cluster distance [18], distance to the BS, and number of neighbor nodes are given to the fuzzy system. All these parameters are denoted by certain levels for example, energy [19] is denoted as low, medium, and high. Depending upon all input parameter values, the rule base for fuzzy logic is prepared. The rule base basically consists of IF–Then Statements. As there are four input parameters to the fuzzy system, and each parameter has three values, i.e., LOW, MEDIUM, and HIGH; therefore, the rule base will contain $81(3^4)$ rules. The Table 1 shows some set of rules in knowledge base of the fuzzy system. Each node will be evaluated based on these rules and the chances of a node becoming a CH will be evaluated. The fuzzy system applies fuzzy rules and output of the algorithm is the probability of a node to become CH [20].

The Table 1 shows the rule base of fuzzy logic system for CH selection in the form of probability of a node becoming CH.

The algorithm used for selecting CH in the network is given below:

Algorithm CH selection

1: Control signal sent by BS to all the sensor nodes
2: Fuzzy Logic applied for CH selection
3: If node i is alive
4: [node i: probability] → [fuzzy_system(node i: energy, node i: neighbours; node i: distance to BS, node i: intra-cluster distance)
5: end if

Table 1 Rule base for fuzzy logic

RE	DBS	ICD	NN	Probability	RE	DBS	ICD	NN	Probability
L	L	L	L	L	H	L	L	L	VL
L	L	L	M	L	H	L	L	M	VL
L	L	L	H	M	H	L	L	H	L
L	L	M	L	M	H	L	M	L	VL
L	L	M	M	M	H	L	M	M	VL
L	L	M	H	L	H	L	M	H	L
L	L	H	L	L	H	L	H	L	VL
L	L	H	M	M	H	L	H	M	VL
L	L	H	H	M	H	L	H	H	L
L	M	L	L	VL	H	M	L	L	M
L	M	L	M	VL	H	M	L	M	M
L	M	L	H	L	H	M	L	H	H
L	M	M	L	VL	H	M	M	L	M
L	M	M	M	L	H	M	M	M	M
L	M	M	H	L	H	M	M	H	H
L	M	H	L	VL	H	M	H	L	M
L	M	H	M	L	H	M	H	M	H
L	M	M	H	L	H	M	M	H	VH
L	H	L	L	VL	H	H	L	L	VL
L	H	L	M	L	H	H	L	M	L
L	H	L	H	L	H	H	L	H	L
L	H	M	L	L	H	H	M	L	L
L	H	M	M	L	H	H	M	M	L
L	H	M	H	L	H	H	M	H	M
L	H	H	L	L	H	H	H	L	M
L	H	H	M	L	H	H	H	M	M
L	H	H	H	M	H	H	H	H	H

```
sort ([node i: probability] in descending order)
for i = 1: i ≤ opt_cluster +1: i++;
for j = 1: j ≤ node_count: j++;
if [node i: probability] < [node j: probability] then
node i ← member_node
end if
end
```

In order to increase the security of the network, the BS uses ELGAMAL algorithm. The BS will apply the algorithm to generate public key and private key of all the

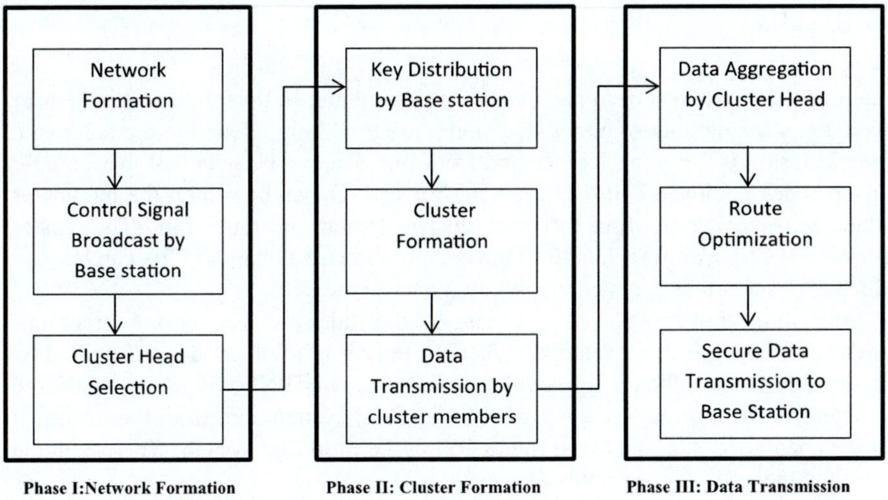

Fig. 2 System architecture

nodes. The private key is transferred to all the nodes before data transmission phase (Fig. 2).

Phase III: Data Transmission

Once the CH is selected, it will send the join message to all the cluster members in its range. The sensor nodes near the CH will send the acknowledgment to the nearby CH. When the clusters are formed then data transmission process will commence. In this phase, the sensor nodes of a cluster send data to the CH and remain in idle state. The CH aggregates [22] the data received from all other nodes and encrypts it using its private key. For deciding route from CHs to the BS, inter-cluster distance is calculated. It is the distance between the elected CHs.

For calculating the inter-cluster distance, the formula used is as follows:

$$\text{Dist\_InterCluster} = \min\left(\|c_i - c_j\|\right)^2,$$

where $i = 1$ to no_c $- 1$ and $j = i + 1$ to no_c and c_i, c_j are centers of CHs and no_c = No. of clusters.

In calculating inter-cluster distance, minimum value is to be considered as route to the BS needs to be optimized. According to the inter-cluster distance, chain of nodes is formed and data is transmitted through this chain to the BS.

3 Results

The proposed system is designed using Jung Simulator. In the initial stage, the user has to enter the number of nodes. The nodes are randomly allotted in the field area. The clustering process takes place and from Fig. 3 it can be seen that there are 40 sensor nodes deployed and 4 clusters are formed. Depending upon the parameter values and fuzzy logic, four CHs are elected depicted in blue color. The cluster members are transmitting data to CH in each round and a chain of CHs can be seen which forwards all aggregated and secured data to the BS.

After some number of rounds, the energy of the nodes is consumed and sometimes it reaches below threshold value. So when the energy of all the nodes reaches below the threshold value, then it is known as Full Node Die (FND). Figure 4 shows the comparison of the proposed system with weighted system and, from the results, it can be concluded that FND is at round 160 in the proposed system, whereas in the weighted method, it is at round 64.

The position of the BS also plays an important role. The energy consumption is calculated by placing the BS at three locations, i.e., inside the network, near the system, and far away from the network. It is clearly seen that, when the BS is inside or near the network, the energy consumption of the network is less as compared to far away by 2–3% (Fig. 5).

Figure 6 shows the energy consumption in weighted and proposed methods. From the graph, it can be seen that the energy consumption is 8% less in fuzzy system as compared to weighted method. So fuzzy system reduces the energy consumption and hence increases the life of the network.

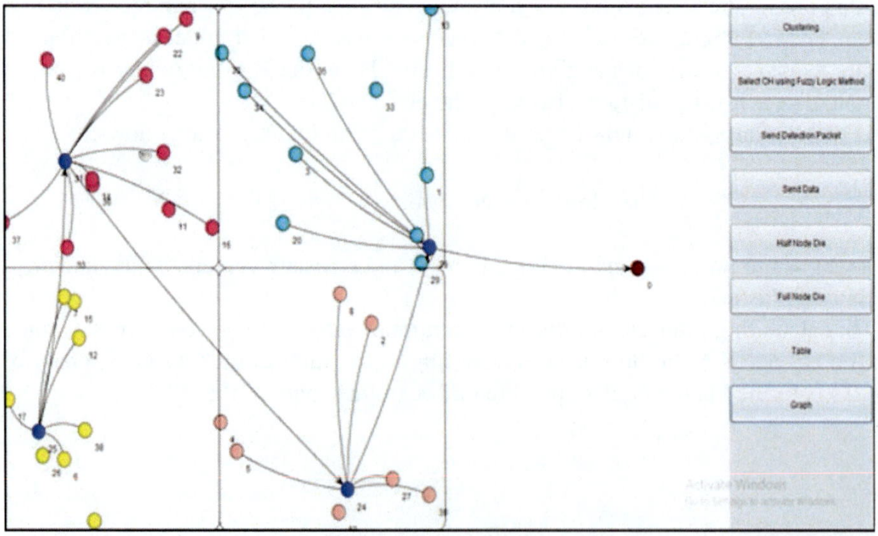

Fig. 3 Cluster head selection and cluster formation

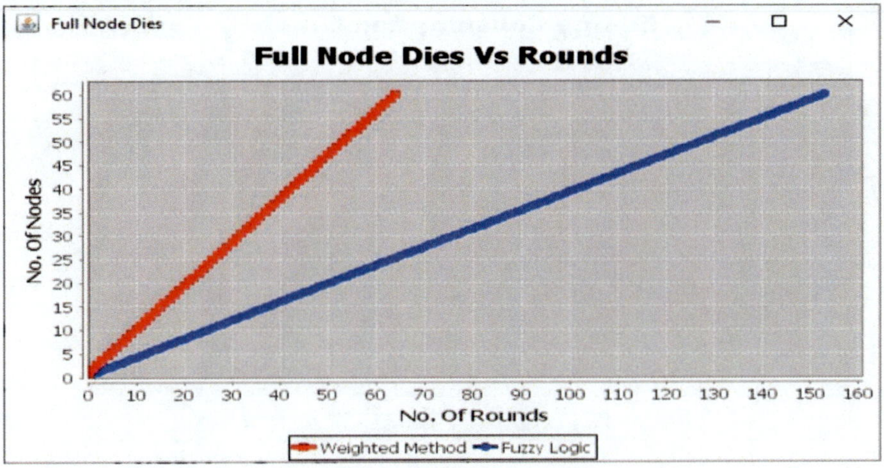

Fig. 4 Full node die (FND) comparison of fuzzy logic and weighted method

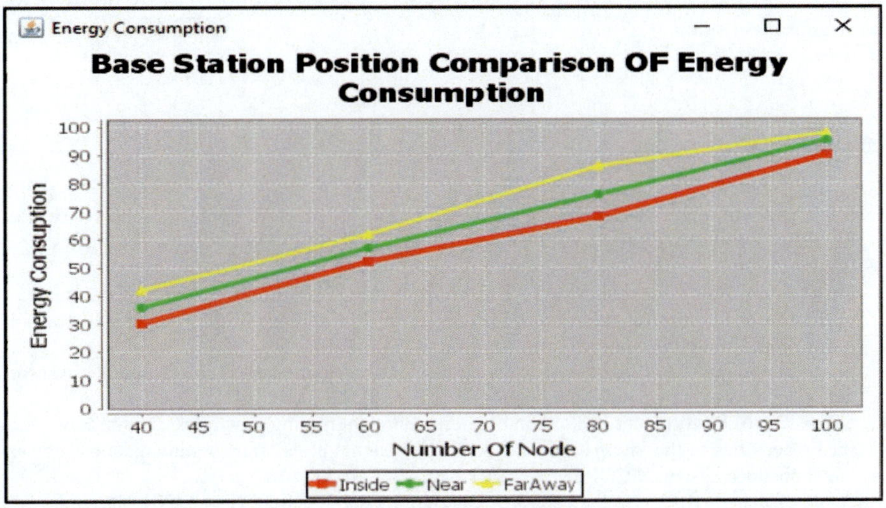

Fig. 5 Energy consumption based on position of base station

4 Conclusion

In this chapter, a centralized unequal clustering method is developed which is energy efficient and secure than the other systems. Due to the use of clustering algorithm based on type-2 fuzzy logic, the energy consumption of the system is reduced. The fuzzy controller is designed based on new parameters, i.e., intra-cluster distance compared from the previous work. The CH selection algorithm has also improved

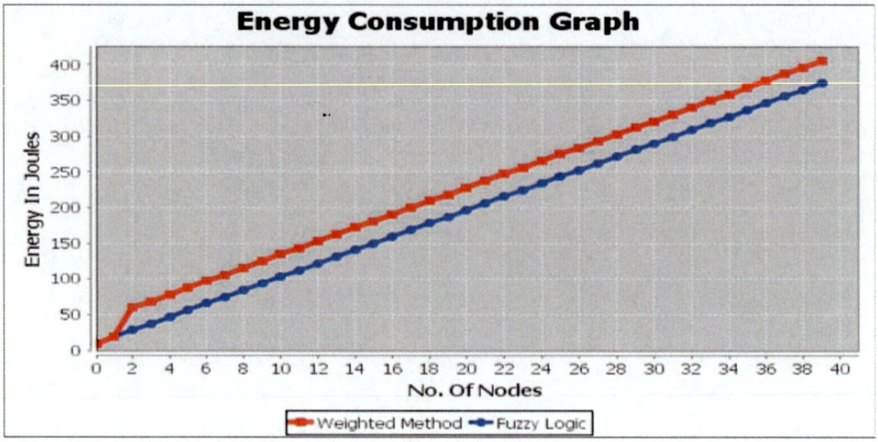

Fig. 6 Energy consumption comparison between weighted and Fuzzy system

the FND and half-node die. As a future scope, the algorithm may be applied to heterogeneous nodes.

References

1. Misra S (2016) A literature survey on various clustering approaches in wireless sensor network. In: IEEE 2nd international conference on communication, control and intelligent systems (CCIS)
2. Heinzelman WR, Chandrakasan A, Balakrishnan H (2000) Energy efficient communication protocol for wireless microsensor networks. In: Proceedings of the 33rd annual Hawaii international conference on system sciences, Jan 2000, p 10
3. Singh SP, Sharma SC (2015) A survey on cluster based routing protocols in wireless sensor networks. Procedia Comput Sci 45:687–695
4. Sah N (2016) Performance evaluation of energy efficient routing in wireless sensor networks. In: Proceedings of the international conference on signal processing, communication, power and embedded system (SCOPES), Oct 2016, pp 1048–1053
5. Asorey-Cacheda R, García-Sánchez A-J, García-Sánchez F, García-Haro J (2017) A survey on non-linear optimization problems in wireless sensor networks. J Netw Comput Appl 82:1–20
6. Agrawal D, Pandey S (2018) FUCA: fuzzy-based unequal clustering algorithm to prolong the lifetime of wireless sensor networks. Int J Commun Syst 31(2):e3448
7. Logambigai R, Kannan A (2016) Fuzzy logic based unequal clustering for wireless sensor networks. Wirel Netw 22(3):945–957
8. Gupta I, Riordan D, Sampalli S (2005) Cluster-head election using fuzzy logic for wireless sensor networks. In: Proceedings of the 3rd annual communication networks and services research conference (CNSR), May 2005, pp 255–260
9. Xie W-X, Zhang Q-Y, Sun Z-M, Zhang F (2015) A clustering routing protocol for WSN based on type-2 fuzzy logic and ant colony optimization. Wirel Pers Commun 84(2):1165–1196
10. Sert SA, Bagci H, Yazici A (2015) MOFCA: multi-objective fuzzy clustering algorithm for wireless sensor networks. Appl Soft Comput 30:151–165
11. Bagci H, Yazici A (2013) An energy aware fuzzy approach to unequal clustering in wireless sensor networks. Appl Soft Comput 13(4):1741–1749

12. Cuevas-Martinez J, Yuste-Delgado AJ, Triviño-Cabrera A (2017) Cluster head enhanced election type-2 fuzzy algorithm for wireless sensor networks. IEEE Commun Lett 21(9):2069–2072
13. Yuste-Delgado AJ, Cuevas-Martinez JC, Triviño-Cabrera A (2019) EUDFC: enhanced unequal distributed type-2 fuzzy clustering algorithm. IEEE Sens J 19(12)
14. Khan A, Tamim I, Ahmed E, Awal MA (2012) Multiple parameter based clustering (MPC): prospective analysis for effective clustering in wireless sensor network (WSN) using K-means algorithm. Wirel Sens Netw 4:18-24
15. Zhang Q-Y (2014) A clustering routing protocol for wireless sensor networks based on type-2 fuzzy logic and ACO. In: 2014 IEEE international conference on fuzzy systems (FUZZ-IEEE)
16. Zhang F (2013) ICT2TSK: an improved clustering algorithm for WSN using a type-2 Takagi-Sugeno-Kang fuzzy logic system. In: 2013 IEEE symposium on wireless technology and applications (ISWTA), September 22–25, Kuching, Malaysia
17. Zhang Y (2017) Network energy efficient based on fuzzy interference system. In: IEEE 29th Chinese control and decision conference (CCDC)
18. Riordan D, Gupta I, Sampalli S (2005) Cluster-head election using fuzzy logic for wireless sensor networks. In: 3rd annual communication networks and services research conference (CNSR'05)
19. Fazackerley S, Paeth A, Lawrence R (2009) Cluster head selection using RF signal strength. In: 2009 IEEE Canadian conference on electrical and computer engineering
20. Bajaber F, Awan I (2009) Centralized dynamic clustering for wireless sensor network. In: International conference on advanced information networking and applications
21. Bajaber F, Awan I (2008) Dynamic/static clustering protocol for wireless sensor network. In: IEEE second UKSIM European symposium on computer modeling and simulation
22. Sunanda VK, Jyothi R (2014) Survey on dynamic clustering for energy efficient data aggregation technique using secure data encoding scheme for WSN. Int J Eng Res Technol (IJERT) 3(2). ISSN: 2278-0181

An Improved K Means Algorithm for Unstructured Data

T. Mathi Murugan and E. Baburaj

Abstract An important significantly applied data mining approach is the clustering. Since the development in technology has increased, rapid generation of large unstructured datasets arises. Traditional clustering does not operate efficiently because of its large processing time. As a result, MapReduce k mean algorithms are developed to process larger unstructured data but it may suffer from disadvantage related to poor quality and inefficiency. In this research, an additional LGC (Local Gravitational Clustering) is integrated with MapReduce k mean known as MapRedLGC module is proposed, which plays a key role in partitioning the data point into clusters on the basis of Euclidean distances. LGC views the data points as mass objects related with produced local resultant force (LRF) by the corresponding neighbours. Significant difference exists in the LRF data points across the centre and boundary, which are analyzed by investigating the centrality (CE) and coordination (CO) factors. The simulation results of proposed method are compared with traditional algorithm in terms of elapsed time, intra-cluster distances and SSE, in which an observed result shows that the proposed technique possess better performances.

Keywords MapRedLGC · K-means · MapReduce · Local gravitational clustering · Centrality · Coordination · Resultant force · Data clustering

1 Introduction

The process of partitioning a structured and unstructured dataset into numerous clusters is known to be the data clustering [1]. The task of data mining relies mainly on extracting an important pattern of knowledge from data set using an unsupervised

T. Mathi Murugan (✉)
Faculty of Computer Science and Engineering, Sathyabama Institute of Science and Technology, Chennai, India
e-mail: mnamashivayam11@gmail.com

E. Baburaj
Department of Computer Science Engineering, Marian Engineering College, Trivandrum, Kerala, India

R. Patgiri et al. (eds.), *Proceedings of International Conference on Big Data, Machine Learning and Applications*, Lecture Notes in Networks and Systems 180,
https://doi.org/10.1007/978-981-33-4788-5_3

data clustering approach [2]. Datasets are classified into numerous clusters, in which the objects belonging to each cluster are similar but different from objects of other clusters [3]. Data relations are discovered using mining approach by following statistical principles of artificial intelligence, database systems, etc. [4]. Clustering documents depend mainly on the repetition of words in the datasets, which can be easily achieved by considering the clustering algorithm as a basic. In addition, clustering groups the documents into cluster in which the properties are found to be similar [5]. Clustering paves a specific way to explore data and to create predictions thus overcoming the inconsistency or deviations in the standard data [6]. Especially in Big Data era, mining data from several sources for the purpose of extracting information are concerned as a challenging one [7]. Data clustering finds applications in several fields like data mining, Pattern recognition, text mining, signal processing, image processing, etc. [8]. The most commonly used algorithm for clustering unstructured data is k-means since it provides simple and highly efficient operations.

K-means is an iterative clustering algorithm that combines N dataset object into k cluster thus achieving higher intra and lower inter distances with respect to similar grouping objects [9]. K-means algorithm also known as nearest centroid classifier or Rocchio algorithm takes the k object as input and determines the Euclidean distance between the object and the centroid. Respective algorithm converges after satisfying prioritized conditions or in completion of finite iterative sequences. Every iteration assigns a group of objects to a nearer centroid that possesses lower distance. Computation of newer centroids is performed on suitable data objects depending on the mean values of every group and the evaluated centroids are provided as input to further iterations. Euclidean distance is considered to be the mostly utilized measure for evaluating the distance [10].

Due to development in technology and increased data volume, the k-means algorithm does not meet the demand. The data sources usually comprise numerous amount of unstructured information. Processing unstructured data using traditional k mean clustering results in larger processing time and poor quality performance. Distributed computing aims to solve these problems by distributing the computations through network comprising interconnected devices [11]. Data mining acts as a framework that processes large datasets in the computer network by means of programming module known as MapReduce that are extensively used in analysis of large-scale data. Such algorithms consist of two-stage processing model known as map and reduce. Parallelization and execution of written programs for determining clusters are performed automatically [12].

Different platform utilizes the successful implementation of MapReduce concept to analyze larger datasets. Mainly, it depends on two functions known as map and reduce in which the output of map function is given as input to reducer. Initially, partitioned dataset is allotted with data chunks and are further provided to reducer to obtain processed output [13]. Integrating MapReduce with gravitational clustering improves the clustering quality through reflecting the similarity existing among data points and its neighbours. Suitable measure is designed to identify the data points possessing at the border clustering regions.

Organization. The paper organization is described in the following manner. Section 2 dictates the represented works relating integrated MapReduce approach along with suitable improved k-means and Gravitational algorithm. Section 3 explains the methodology background. Section 4 depicts the reason to select the proposed method. Section 5 describes the design of newly proposed MapReduce k mean algorithm and local gravitation model referred as MapRedLGC which are useful for clustering. Section 6 evaluates the simulated results and performance analysis of the proposed integrated algorithm. Section 7 illustrates the conclusions and the adopted works to be analyzed in future.

2 Literature Review

Efficient MapReduce implementation of k mean uses map and reduces functions in clustering larger texts dataset across different interconnected nodes. Larger dataset clustering seems more efficient than smaller [14]. Similarly, K-means algorithm with MapReduce programming model uses combiner between mapper and reducer. Combiner enhances the model performances by reducing the amount of write and read in the mapper and reducer, respectively, but does not concentrate on reducing the redundant for smaller datasets as well as single point failure in cluster [15]. Dividing and aggregating the tasks of mining algorithm such as k-means, genetic k-means and PSO by means of Map and Reduce process like Parallel Metaheuristic Data Clustering Framework (PMDCF) is proposed on cloud computing environment to minimize the run time and solving complexities. Analyzing mining procedures and internet of things (IoT) data needs to be developed [16]. Several evolutionary techniques are introduced through the utilization of simple MapReduce approach to solve computationally intensive drawbacks [17]. Hybridized k-means-PSO with MapReduce technique [18] and MapReduce dependent artificial bee colony algorithm (MR-ABC) to cluster complex larger datasets [19] is presented. Similarly, frequency-based k-means Bat technique is adopted to vary the bat frequency dynamically for improving the accuracy and handling larger datasets [20].

K-means with MapReduce clusters text documents by transforming the vector models, which is then provided as input to the masters. Master node comprises random selected vector centroid values and is given as input to slaves. The mapper evaluates the centroid by computing minimized distance between the clusters and further the reducer recomputes the centroid [21]. Modified k-means with MapReduce uses pruning-based approach at the mapper known to be triangle inequality for achieving reduced computational distance. As the dataset increases, speed up produces linear output. For larger datasets, it produces efficient output [22]. Parallel k-medoids with MapReduce is accomplished by choosing randomized objects. The data object distances are computed based on the medoids and the outcomes are provided to the reducer thus obtaining reduced communicational overhead. As k-medoid possesses large execution time, it does not operate well for complex datasets [23].

Gravitational search algorithm (GSA) improves the performance of k-NN (Nearest Neighbour) to provide better accuracy solutions. Instead vote rule is also used for further improving the behaviour of classifier. This option is applicable only for prototype generation not suitable for prototype selection hence serves as a drawback for future evaluations [24]. Every data point object is assigned with mass and local resultant force (LRF) and significant variation exists between data points at the centre and its boundaries. Therefore, investigation of two measures like CE and CO is performed. Depending on the empirically observed results, clustering approaches such as local agent communication and LGC is implemented for verifying its effective utilities. Better performance is observed with enhanced quality [25]. Newly categorized k-means algorithm known as Manhattan Frequency k-Means (MFk-M) transforms data into numerical values using the frequency attribute. Manhattan distance is evaluated as the distance between the objects and centroids. The advantage of this approach is high efficiency and reduced complexity. Further its efficiency can be improved by integrating with genetic algorithm, cluster centre initialization, etc. [26].

Integrating k-means with data transformation model avoids generating empty cluster hence providing better accuracy and optimal solution for real datasets than k-means algorithm. In addition to this, modified silhouette algorithm evaluates the cluster number that gains an added advantage. Time-consuming operates faster when compared with k-means and k-means++ but results in complex structure regarding the integration [27]. Hybrid model-based and hierarchical-based clustering data mining algorithm that transforms the Markov model into probabilistic space and sequence clustering, respectively, thus enable its purpose in guiding as well as social security policies. Extensions based on computing distance using probability measure can be adopted to explore the categorical data sets [28]. BFGSA (Birds Flock GSA), a newly developed GSA diversity mechanism that inspires the bird behavioural responses collectively. Furthermore, three different steps are involved in enhancing the mechanisms such as initialization process, nearest neighbours identification and orientation variations. Updating positional values solves the premature convergence problems in BFGSA by rapidly reducing the diversities. High performance is achieved in terms of error rate [29].

3 Background Methodology

MapReduce K-means: Parallel k-means is integrated with MapReduce to cluster text documents and the comparison between the present technique and sequential k-means with respect to execution time is done for different sized datasets. The simulation results indicate the dataset clustering within shorter time period through utilization of 10 nodes. Incorporating mapper and reducer effectively improves the performance. This approach suitably operates well for larger datasets than smaller.

Local Gravitation clustering (LGC): The relation among data points and its neighbours are analyzed based on the effective local resultant field (LRF). New approaches are designed by formulating appropriate measures for addressing drawbacks in analysing cluster. In this, clustering by local agents as well as balanced strategy is adopted to obtain improved clustering quality. Also it operates based on two parametric factors like CE and CO.

4 Motivation of the Proposed Work

From the literature studies, each algorithm offers advantages and disadvantages. The efficiency and quality of integrated k mean algorithm are further suggested to improve. Generally, gravitational clustering algorithm provides better performance by clustering the data points. Hence there exists an alternative choice to combine both MapReduce k mean with LGC to meet the requirements of existing k mean approaches.

The main aim of this research is to integrate MapReduce k mean with LGC to show the relational linkage among the data points and its neighbours to offer better quality with efficient operation. As a result of this, MapRedLGC technique came into existence. The cluster analysis partitions data point sets into numerous groups with appropriate suitable distances. According to LGC, design of particularly two local measure known as centrality (CE) and coordination (CO) is performed for the purpose of identifying data points that present in border cluster regions. LGC works on the basis of Newton's gravitational theories in which data points act as mass objects and the LRF is associated with every data points by the neighbours. The LRF magnitude and directions are taken into considerations.

5 Proposed MapRedLGC Algorithm

In this proposed work, traditional k-means clustering data mining algorithm is transformed to parallel k-means by incorporating MapReduce programming due to dissatisfying the demand of larger datasets and developments in technology. The main aim of this research is to improve the cluster efficiency and quality based on elapsed time with larger documents. Also MapReduce performs document clustering of unstructured data using vector space model. But integrating these models does not solve the issues of k-means that depends or concentrates only on the selection of centroid between objects of datasets. Further modifications need to be done for obtaining better quality with improved efficiency. One such option for satisfying these requirements is to integrate MapReduce k-means with local gravitation clustering.

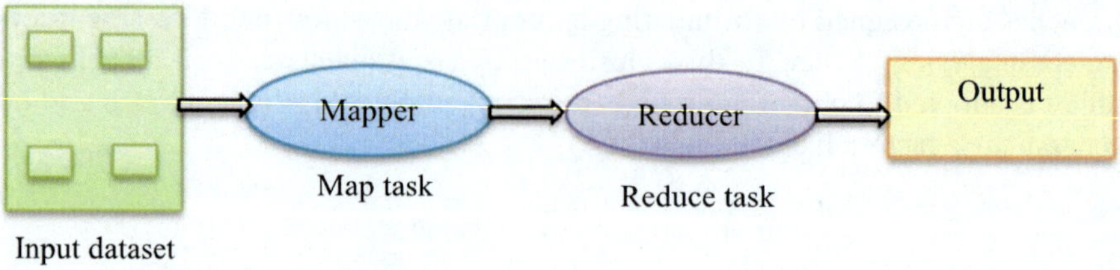

Fig. 1 Processing scheme of MapReduce

5.1 MapReduce Programming Module

A set of <key, value> input is given to MapReduce to produce pair set <key, value> as output. Two functions such as Map and Reduce are performed by this programming module in k-means mining algorithm. Initially, the input data sets are transformed to <key, value> before passing the data to mapper. Mapper analyzes the data with keys and values.

a. **Map function**. It accepts the input datasets to produce an intermediate pairs. Each key is associated with a specified value. The input data are read by mappers using distributed system and generates the results, which are stored in file system. Based on similar keys, the intermediate results are grouped together. The location of this result is provided to the reducer so that results can be easily accessed.
b. **Reduce function**. Reads the data from mapper and applies the reduce function thus output the results on the basis of key value pair. The data processing scheme of MapReduce programming is depicted in Fig. 1.

5.2 K-Means-MapReduce Algorithm

The steps followed in processing large volume data sets using k-means MapReduce model are as follows.

Step 1: Processing large volume data sets: The input text documents are transformed or processed to a form that are suitable for mapping function regarding input of k-means algorithm. Numerous methods like vector models, stemming, graphical analysing models, etc. are involved for preprocessing input datasets. Since k-means takes numeric input datasets, vector model technique is preferred for preprocessing that specify words in datasets into numeric quantity depending on its occurrence and is saved in a file, which is distributed to k-means as an input.

The multi-dimensional numeric vector values comprise multiple dimensions, which represents the weight of each word. Each terms in input datasets t_i, where $i = 1, \ldots, n$ corresponds to vector quantity: d that corresponds to different element sets $\{w_1, w_2, \ldots, w_n\}$, where w_n links with t_i. Preprocessing steps involved in vector space model involve text normalization, removal of smaller or higher frequencies, stop words removal.

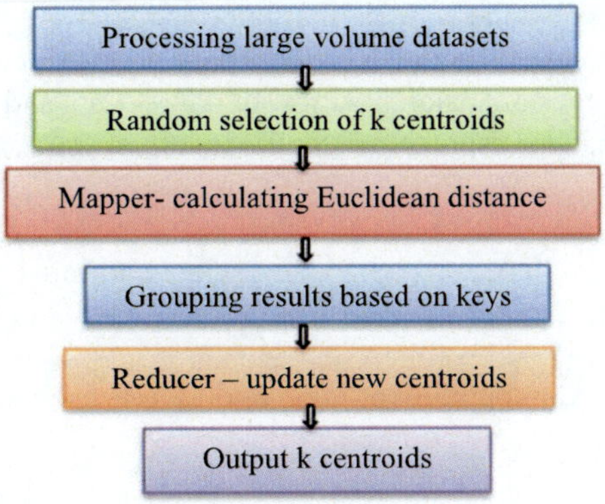

Fig. 2 Schematic execution process of MapReduce K-means

Step 2: K-cluster centroid Specification: Before execution of the algorithm using MapReduce, k-means randomly generates the cluster centroid k and accumulates in a centroid file and finally supplies it to the map function. Keeping the k value as same further map function iteratively recalculates distance between the objects and centroid and updates them.

Step 3: Implementation of MapReduce k-means programming modules: After the evaluation of the centroid by mapper, reducer further determines the new centroid based on sum and number of objects thus outputs the old and new centroid respectively. Also this technique determines the elapsed time of the programming model.

K-means algorithm execution is partitioned into two parts. First, evaluating Euclidean distance between objects and centroid iteratively is done using respective equation and then assigning nearest centroid to the objects, respectively. This process of assigning objects is referred as the map function. Second, while performing reducer function, new centroids are updated sequentially after every iteration. At last, the old and new centroids evaluated by map and reduce functions are displayed depending on the cluster numbers. Figure 2 shows the overall execution process of k-means.

5.3 Local Gravitational Clustering

LGC algorithm evaluates the LRF of every data points and using this parameter, clustering tasks are performed based on its classification as interior, unlabelled and boundary points, and further soft-connecting process is done on the interior data points whereas cluster assignment is performed by the boundary data points. Enhanced LGC known as balanced strategy (LGC-B) is integrated with MapReduce k-mean to offer improved clustering quality. Such enhanced algorithm searches the data points with smaller LRF and larger CE in k mean clustering. Key advantage of utilising LRF depends on the fact that relative smaller force exists on interior region whereas the force tends to be large in boundary regions. Other properties of LRF involve the concept that interior region possess random direction of data points but in case of boundary areas, data points direct towards its centres. Initially, resultant forces are calculated for all data points and then index k is computed using Eq. (1) that corresponds to smaller resultant force.

$$\mathrm{k}_i = \underset{k}{\operatorname{argmin}} \left(\left| \sum_{j=1}^{2k} D_{ij} \right| \right) \tag{1}$$

Computation of mass and LRF is evaluated using Eqs. (2) and (3).

$$m_i = 1 \Big/ \sum_{j=1}^{k_i} D_{ij} \tag{2}$$

$$F_i = \frac{1}{m_i} \sum_{j=1}^{k_i} D_{ij} \tag{3}$$

Balanced strategy techniques are applied in CE and CO evaluations by Eqs. (4) and (5) in which CE calculations require numerous data points.

$$CE_i = \frac{\sum_{j=1}^{k_i'} \cos\left(\overrightarrow{F_j}, \overrightarrow{D_{ji}}\right)}{k_i'} \tag{4}$$

$$CO = \sum_{j=1}^{k_i'} \left(\overrightarrow{F_i} \cdot \overrightarrow{F_j}\right) \tag{5}$$

The number of k mean neighbour to evaluate CE and CO is specified below in Eq. (6).

$$\phi(m)_i = \sum_{j=1}^{m} \left| F_{ij} \right| \tag{6}$$

In which m indicates the variable with lower values in comparison to overall data points whereas $\phi(m)$ represents the magnitude sum of nearest neighbour of k mean clustering. After LRF computations, data points are sorted based on the magnitude of resultant force and transformed further to SortMagnitude vectors. Magnitude of threshold forces is indicated below in Eq. (7)

$$\psi = k \cdot SortMagnitude(\lceil cn \rceil) \tag{7}$$

where k denotes the number of neighbours, c varies between the constant values 0 and 1 and n indicates the data point quantity. Here the value of c is set to be 0.6.

6 Results and Discussions

The experiments are performed on text documents by assuming the cluster number as $k = 8$. Larger unstructured weather datasets that contain detailed description related to damage caused, locations, duration, etc. are used in these experiments. For larger text datasets, preprocessing plays a significant role, which involves the stop words removal, normalizing etc. After obtaining preprocessed text documents, the centroid is generated randomly using k mean clustering algorithm in which k is set to be 8. The clustering simulated results of k means are shown in Fig. 3.

The randomly generated cluster centroid and the pre-processed vector are given as key value pair to the input map function. Mapper recalculates the cluster centroid using the Euclidean distance between the objects and the centroids. Further reducer iteratively calculates the centroid and outputs the old as well as new centroids, respectively. The evaluated value of MapReduce is shown in Table 1 for $k = 8$.

LGC clustering algorithm is applied on the dataset with parameter setting of MapReduce k mean to partition the data automatically into different cluster groups.

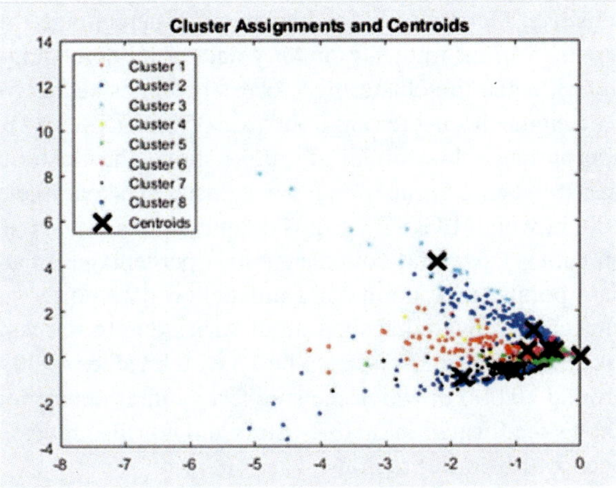

Fig. 3 Clustering output of k mean algorithm

Table 1 MapReduce output with respect to centroid calculation

Old centroids	New centroids
−1.4012	−2.5469
−1.2391	−1.2465
−0.9959	−0.9044
−0.7151	−6.9547
−0.7102	−0.9928
−0.7021	−1.7354
−0.4760	−3.3578
0	−1.2479

Fig. 4 Partitioning data points into clusters using LGC

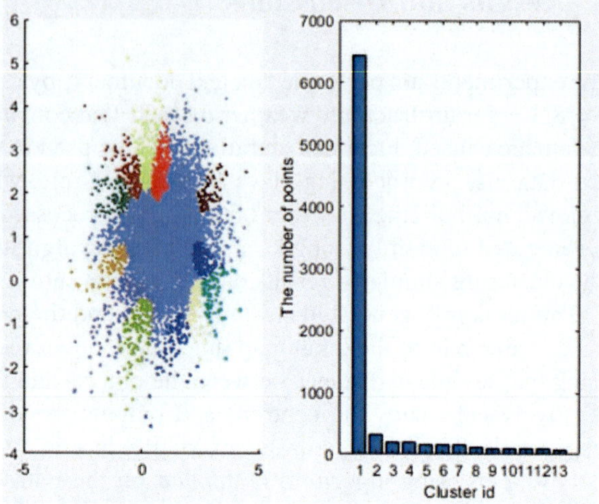

LRF relies mainly on the size of neighbourhood but the dependency is not much sensitive. Generally k lies between the percentage ranges 0.5–1.5% of entire data points. Similar to this, boundary data point percentage influences the critical value of CE hence the clustering process becomes successful when boundary data point percentage lies in the range between 35 and 75%, and this is accomplished by setting appropriate values for CE. Figure 4 shows the clustering results of LGC-B in which data points are grouped to form a cluster. The parameter IM is varied between 0 and 400, in which LGC-B assigns default values to the parameters k and CE depending on data set size and boundary points percentage. In addition, the number of border data points, core points and unlabelled data points are evaluated. The number of clusters discovered at the output varies due to the variations in the neighbourhood size. In this case, IM is set to be 10 with k values of 30, total number of data points is around 10,000 in which the smallest value is determined to be 78. The output result shows the formation of 13 clusters and it is also revealed that most of the data points occupy the border regions.

6.1 Performance Analysis

A. **Intra cluster distances**. Defined as the distance that exists between data vectors and the cluster centroids as given in Eq. (8). Lesser the value of intra-cluster sum distances, higher will be the clustering quality.

$$D(x_p - z_j) = \sqrt{\sum_{i=1}^{d} (x_{pi} - z_{ji})^2} \tag{8}$$

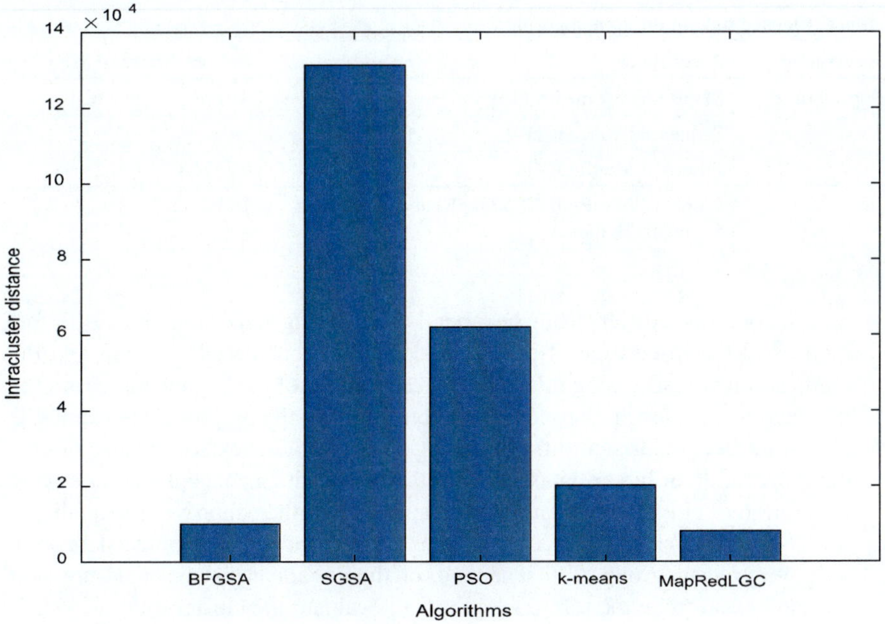

Fig. 5 Intra-cluster distance of different algorithms

In which z_j specifies the centroids of cluster j; x_p indicates the pth data vectors; d dictates the feature quantity of every cluster centroids. Figure 5 shows the evaluated intra-cluster distance of different clustering algorithms. Average value of intra-cluster sum distances is reported. The proposed MapRedLGC algorithm is implemented on 250 MB datasets and the observed results are compared with K-means [30], standard version of GSA [31], BFGSA [29], PSO [32].

From the observed simulation results, the intra-cluster distance obtained by the proposed MapRedLGC offers decreased value than that of the other methods like BFGSA, SGSA, PSO and K-means. Decreased intra-cluster distance indicates a higher clustering quality. Among the existing techniques, BFSA produces smaller intracluster distance and SGSA offer larger intracluster distance. Decreased value of distance is obtained due to the integration of MapReduce concept with local gravitational clustering.

B. **Elapsed time**. Table 2 shows the observed elapsed time of numerous algorithms in related to dataset sizes. This parameter acts as a framework in analysing and evaluating the performances of proposed and existing k-means algorithms.

Sequential k mean consumes an elapsed time of 30.2 s for 250 MB datasets whereas MapReduce k mean takes 10.2 s and the proposed k-means took 0.8776 s, respectively. Investigation of elapsed time is performed in order to analyse and the performance comparison of different algorithms is presented. Ratio acts as an efficient source to compare the analysed performances. Ratio signifies the relation among two different quantities, which predicts the number

Table 2 Elapsed time of different algorithm

Observation	Algorithm	Dataset size (250 MB)
Elapsed time	MapReduce k means [14]	87 s
	Sequential k means [14]	195 s
	Proposed MapRedLGC	0.8776 s
Ratios	Proposed MapRedLGC: MapReduce k means: Sequential k means	1:99:222

of times one value differs from the other. For analysing and evaluating the performances, the elapsed time of proposed k mean is considered as a unit and the comparison is made using different representatives of other k means. From the obtained results, lower elapsed time is consumed by the proposed MapRedLGC algorithm in comparison to traditional k means and this exists because of efficient operation of larger datasets using the MapReduce concept and automatic generation of cluster by updating mass and LRF further improves the quality.

C. **Sum Squared Error (SSE)**. The k-means clustering relates the object data point x_i to every cluster groups C_i that depends on the Euclidean distance measurement m_i and is said to be the reference points to evaluate the cluster quality. SSE is known to be one of the approaches in validating the clustering technique. It can be used as a measure of variation within a cluster as depicted in Eq. (9). If all cases within a cluster are identical the SSE would then be equal to 0.

$$SSE(X) = \sum_{i=1}^{k} \sum_{x_j \in C_i} \|x_i - m_i\|^2 \tag{9}$$

where x denotes the objects and m specifies the cluster centroid. Figure 6 shows the comparison of different algorithms in terms of SSE. The proposed k mean exhibits smaller error rate in comparison to other approaches as depicted thus outperforms the existing techniques like BFGSA, k-means and GSA with respect to quality and efficiency. Table 3 and Fig. 6 present the SSE comparison of different algorithms and their simulation results.

In case of k-means and GSA, increased error rate exists that represents the frequent occurrence of clustering quality degradation. When comparing the BFGSA and proposed MapRedLGC, the proposed technique provides smaller error rate and the reduction of this parameter automatically increases the quality by providing efficient operation. Next to MapRedLGC, BFGSA produces smaller error rate with improved clustering quality.

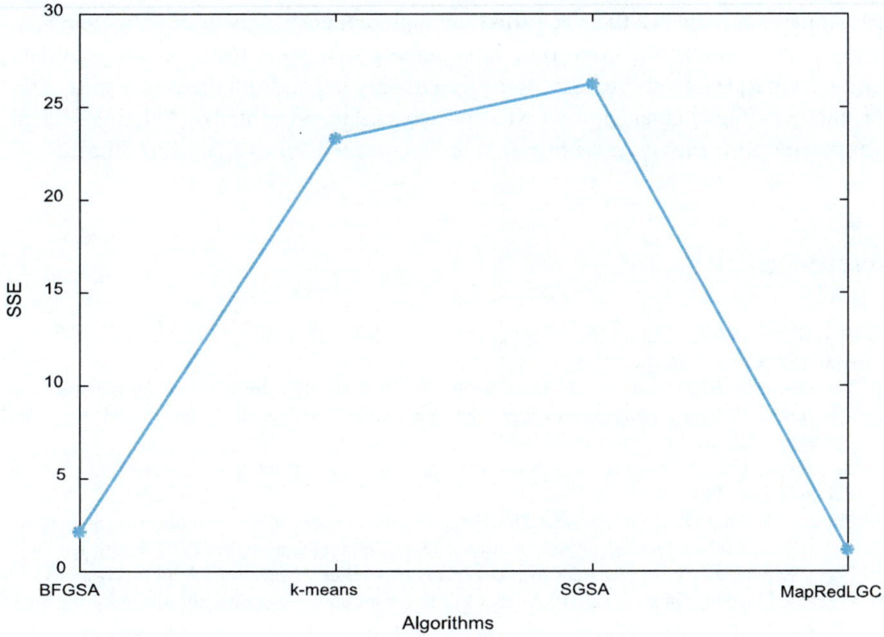

Fig. 6 Simulation results of SSE

Table 3 SSE comparison of various algorithms

Methods	SSE
BFGSA [29]	2.14
K-means [30]	23.31
SGSA [31]	26.26
Proposed MapRedLGC	1.14

7 Conclusion

MapReduce k mean modules analyze larger datasets within a short time period. The quality and efficiency of this integrated technique are further enhanced by incorporating LGC. Clustering by local gravitation analyzes the relation between data points and their neighbourhoods using modelled LRF. Performing clustering task on the basis of directional information is considered as a satisfactory one. The adopted measure and strategy of proposed LGC operate efficiently with larger unstructured data and the obtained results motivate the researchers to model modules further by satisfying the considerations. The proposed LGC MapReduce k mean algorithm known to be MapRedLGC is designed by partitioning data into clusters and the simulated performance results are compared with other similar algorithms. Thus the comparison indicates that the proposed approach possess better quality and efficiency with minimized elapsed time thus making it suitable for application in data mining techniques. However, this algorithm mainly depends on the centre initialization thus gets stuck in local optima. Such drawbacks are prevented by utilizing several hybridized optimization algorithms that are recommended to analyze in future.

References

1. Shyr SY, Shao WC, Chuin MW, Yung KC, Ting CC (2018) Two improved k-means algorithms. Appl Soft Comput 68:747–755
2. Tanachapong W, Sirapat C, Khamron S (2017) Efficient algorithms based on the k-means and chaotic league championship algorithm for numeric, categorical, and mixed-type data clustering. Expert Syst Appl 90:146–147
3. Berikov V (2014) Weighted ensemble of algorithms for complex data clustering. Pattern Recognit Lett 38:99–106
4. Prajesh PA, Anjan KK, Srinath NK (2016) MapReduce design of K-means clustering algorithm. In: International conference on information science and applications (ICISA), South Korea
5. Neepa S, Mahajan S (2012) Document clustering: a detailed review. Int J Appl Inf Syst 4(5):30
6. Preeti A, Deepali, Shipra V (2016) Analysis of K-means and K-medoids algorithm for big data. Procedia Comput Sci 78:507–512
7. Ruili W, Wanting J, Mingzhe L, Xun W, Jian W, Song D, Suying G, Chang-an Y (2018) Review on mining data from multiple data sources. Pattern Recogn Lett 109:120–128
8. Everitt BS, Landau S, Leese M, Stahl D (2011) Cluster analysis, 5th ed. Wiley series in probability and statistics
9. Zhang J, Gongqing W, Xuegang H, Shiying L, Shuilong H (2013) A parallel clustering algorithm with MPI-MK-means. J Comput 8(1):10
10. Bawane SV, Kale MS (2015) Clustering algorithms in MapReduce: a review. Int J Comput Appl 975:8887
11. Tanvir HS, Ahmed RF, Zahid A (2018) An evaluation of MapReduce framework in cluster analysis. In: International conference on intelligent computing, instrumentation and control technologies (ICICICT). IEEE, India
12. Daria G, Petar J, Alberto A (2017) MapReduce performance model for Hadoop 2.x. Inf Syst 79:32–43
13. Tanvir HS, Zahid A (2018) Partition based clustering of large datasets using MapReduce framework: an analysis of recent themes and directions. Future Comput Inform J 3(2):247–261
14. Tanvir HS, Zahid A (2018) An analysis of MapReduce efficiency in document clustering using parallel K-means algorithm. Future Comput Inform J 3(2):200–209
15. Prajesh PA (2014) Improved MapReduce k-means clustering algorithm with combiner. In: 16th international conference on computer modelling and simulation. IEEE, UK, pp 386–391
16. Chun WT, Shi JL, Yi CW (2018) A parallel metaheuristic data clustering framework for cloud. J Parallel Distrib Comput 116:39–49
17. Gong YJ, Chen WN, Zhan ZH, Zhang J, Li Y, Zhang Q, Li JJ (2015) Distributed evolutionary algorithms and their models: a survey of the state-of-the-art. Appl Soft Comput 34:286–300
18. Wang J, Yuan D, Jiang M (2012) Parallel k-pso based on mapreduce. In: 14th international conference on communication technology ICCT. IEEE, China, pp 1203–1208
19. Banharnsakun A (2017) A mapreduce-based artificial bee colony for large-scale data clustering. Pattern Recogn Lett 93:78–84
20. Tripathi AK, Sharma K, Bala M (2017) Dynamic frequency based parallel k-bat algorithm for massive data clustering (DFBPKBA). Int J Syst Assur Eng Manag 9(4):866–874

21. Anchalia PP, Koundinya AK, Srinath NK (2013) MapReduce design of K-means clustering algorithm. In: International conference on information science and applications (ICISA). IEEE, South Korea, pp 1–5
22. Haj Kacem MAB, N'cir CEB, Essoussi N (2016) An accelerated MapReduce-based K-prototypes for big data. In: Milazzo P, Varró D, Wimmer M (eds) Software technologies: applications and foundations, STAF 2016. Lecture notes in computer science, vol 9946. Springer, Cham
23. Srinivasulu DL, Reddy AV, Akula VSG (2015) Improving the scalability and efficiency of K-medoids by MapReduce. Int J Eng Appl Sci (IJEAS) 88:2–4
24. Mohadese R, Hossein N (2015) Using gravitational search algorithm in prototype generation for nearest neighbor classification. Neurocomputing 157:256–263
25. Zhiqiang W, Zhiwen Y, Philip CLC, Jane Y, Tianlong G, Hau-San W, Jun Z (2018) Clustering by local gravitation. IEEE Trans Cybern 48(5):1383–1396
26. Semeh BS, Sami N, Zied C (2018) A fast and effective partitional clustering algorithm for large categorical datasets using a k-means based approach. Comput Electr Eng 68:463–483
27. Rasool A, Mohadeseh G, Mahmoud G, Hedieh S (2017) A novel clustering algorithm based on data transformation approaches. Expert Syst Appl 76:59–70
28. Luca DA, Jose GD (2014) Mining categorical sequences from data using a hybrid clustering method. Eur J Oper Res 234(3):720–730
29. Xiao HH, Long Q, Xiao YX, Matt A, Jie X, Yuan, L (2017) A novel data clustering algorithm based on modified gravitational search algorithm. Eng Appl Artif Intell 61:1–7
30. Nanda SJ, Panda G (2014) A survey on nature inspired meta-heuristic algorithms for partitional clustering. Swarm Evolut Comput 16:1–18
31. Bahrololoum A, Nezamabadi-pour H, Bahrololoum H, Saeed M (2012) A prototype classifier based on gravitational search algorithm. Appl Soft Comput 12:819–825
32. De-Falco I, Della CA, Tarantino E (2007) Facing classification problems with particle swarm optimization. Appl Soft Comput 7(3):652–658

Vision-Based Smart Shot for Assisting Shooters

Joyeeta Singha and Aman Kumar

Abstract Shooting training system can help the shooters to enhance their performance. The objective of the proposed system is to design a cost-effective and environment friendly model. The system has two lasers, which are attached to the barrel of the gun and the camera. One laser is used to track the movement and another to detect the shot. The camera is used to record the training period of the shooter. After preprocessing, the shot is detected and the target is identified. The model gives the analysis of the shot by determining its position, calculating the score, tracking the movement of the gun and the time spent on the target in the different areas. The whole analysis is being plotted on the digital target, which is being formed according to the ISSF rules. Codes have been made publicly available at https://github.com/aman-kumar-jh/Smart-Shot.

Keywords Shooting training system · Laser · Target · Shot

1 Introduction

The shooting training system is the smart model, which improves the performance of the shooter. The system can study the behavior of the shooter by interpreting the movement and stability of the gun around the target, obtaining the position of the bullet and determining the score [8, 10]. The model is being executed using lasers attached to the barrel of the gun [3, 4] and a charge-coupled device sensor camera to capture the video. This model can help international shooters to improve their performance and the defense system of the country can also be benefited from this model.

The proposed idea of the design is not to use a real bullet instead use a laser as a bullet, which is attached to the barrel of the gun. The method to do so uses two lasers. One to track the movement of the laser that will always be activated and another laser to detect the shot that gets initiated only when the shooter pulls the trigger. Tracking

J. Singha · A. Kumar (✉)
LNMIIT, Jaipur 302031, India
e-mail: akumar.jhjh@gmail.com

R. Patgiri et al. (eds.), *Proceedings of International Conference on Big Data, Machine Learning and Applications*, Lecture Notes in Networks and Systems 180,
https://doi.org/10.1007/978-981-33-4788-5_4

of the laser is being done by calculating the slope between every two points. Shot management is also being processed, which indicated the shooter when to be in a ready state, fire state or cancel the shot. The formula used for determining the score is being derived by analyzing the demand and constraint given by the International Shooting Sports Federation (ISSF) [6]. All the processing is being shown on the digital target according to the constraint given by ISSF [6]. The accuracy of the model will profoundly depend on the position of the camera and the composition of the digital target with respect to the target used in the video. So, once the correct position of the camera is known then the highly realistic result can be obtained as the reference target eternally remains fixed. The data set is made in shooting range while maintaining the real scenario.

The paper is organized in the following way. Literature survey is described in Sect. 2, proposed system is presented in Sect. 3, results along with discussion are elaborated in Sect. 4, conclusion along with future work is shown in Sect. 5.

2 Literature Survey

Ding et al. proposed the method to build a computer-based system to derive the shot score by the shooter with the help of optical sensors and man–machine interface. The system uses different image processing techniques to find the shot accurately [4, 9]. The system after finding the shot uses the Circular Hough transform to find the center of the shot [3, 5, 9].

Softendo et al. proposed a computer vision-based system [9] where the laser is attached on a weapon and in the place of bullet hole a laser spot is present and the score gets detected by detecting the laser spot [4, 5]. The system utilizes a camera to record the spot and to calculate the score. It first tries to locate the center of the circular target by using different image processing techniques [5, 9]. The information gained by the above techniques is used to determine the shooting score of the laser spot. The score had been calculated by comparing the distance between the laser spot and the center [4]. The camera is put inside a nontransparent box and the laser used in the setup is a super white bright LED.

Rongxin Du proposed the method to develop a system that will tell the position of bullet holes on a targeted computer near the shooter connected to the system [5]. The detection work on the time difference of arrival (TDOA) of acoustic energy from the supersonic shock wave generated by the bullet as it passes through the target. This type of design used on target is called Soft target design. Alternatively, if shock wave is generated by the target with the impact of the bullet on it is called the Hard target design. Both target designs are used for detection when the shot is fired. Both target type has their benefit, as well as their problems such as soft target only, work on supersonic ammunition hence it typically needs a noise-isolation environment, as well as the target, made up of noise-isolation material to eliminate extra noises. Whereas a hard target required a material, which evenly spread noise over the whole surface of the target to work properly.

Sumit A et al. proposed the shooting training model in which laser is added on the weapon below the barrel and the bullet holes are replaced by the laser spot [2, 4]. The system can track both the target and the laser spot. The camera is connected to the weapon itself the camera rolls with respect to the weapon itself hence the position of shot changes with respect to the hand movement. The system locates the laser spot inside the circular target [5]. When the shooter presses the trigger, a laser beam is fired, which is used to detect the shot score by the system.

3 Proposed System

Figure 1 displays the structure of the model. The laser is attached to the barrel of the gun and a camera is placed away from the target such that it does not obscure the laser vision. The camera records training period of the shooter. Then the data are being analyzed.

Figure 2 represents the preprocessing of the data. The examination of data consists of four vital steps. The process includes finding the target, detection of the lasers, analysis of shot, and building of a digital target. Every step is elaborated in further subsections.

A background image is taken from the data on which image preprocessing is being executed to find the center of the target. The preprocessing involves the manipulation of the original image to get the desired result which will be explained in Sect. 3.1. With the help of the center, the target image is formed. Then the whole data are processed to find the position of the laser and to differentiate between the shot and the tracking. The processing is being performed on the result of the preprocessed data, which are further explained in Sect. 3.2. Once the shot and track of the laser are being detected, then the result of the track is further processed to get an analysis

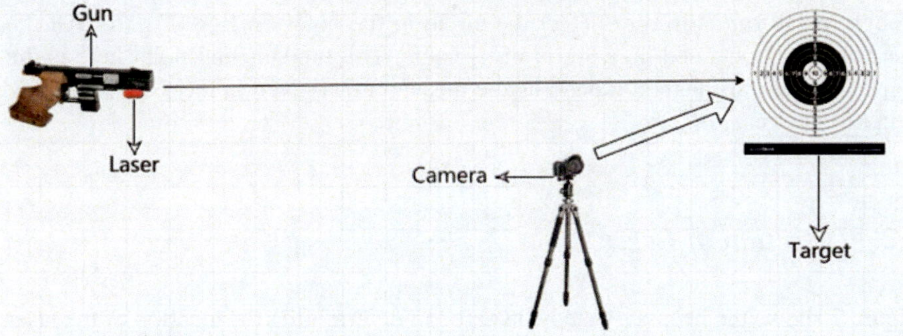

Fig. 1 Setup of the model

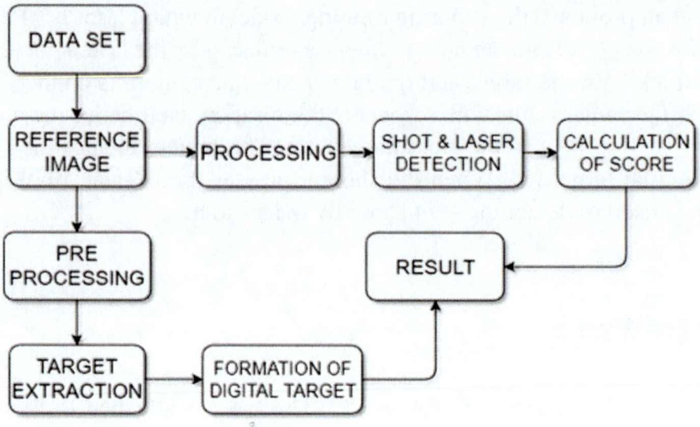

Fig. 2 Flow diagram

of the data. After the process of analysis, a new target is built, and the results are merged with the newly constructed target.

3.1 Finding the Target

The object is to find the center and radius of the target and then just find the offset point of the target according to ISSF rule book measurement. The first frame of the video is taken as a reference image or background image and all the process is executed on a reference image. The process starts with resizing the image to speed up the process (Fig. 3).

The notion behind using the binary image is that the image has a large black circle so if that circle is detected then the desired result can be achieved (Fig. 4). To get the binary image, the image needs to be transformed into gray scale (Fig. 5). Circular Hough transform is used to detect the circle in the binary image (Fig. 6). All the measurements obtained here are in pixel value. The fundamental point here is the conversion of millimeters to pixel as the measurement given in the ISSF rule book is in millimetres.

3.2 Detection of Laser

Figure 7 shows the necessary steps involved in determining the location of the laser in the image. The result of Sect. 3.1 is used in cropping the image. The laser used in our proposed system is of red color. The red component (Fig. 8) from the colored image is subtracted from the gray scale image to obtain a binary image as shown in

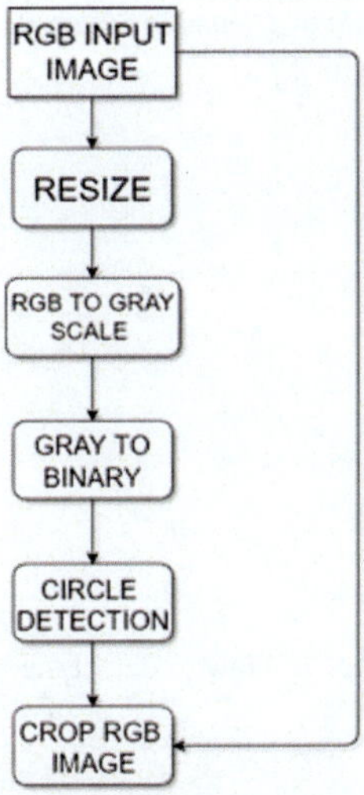

Fig. 3 Flow to detect the target

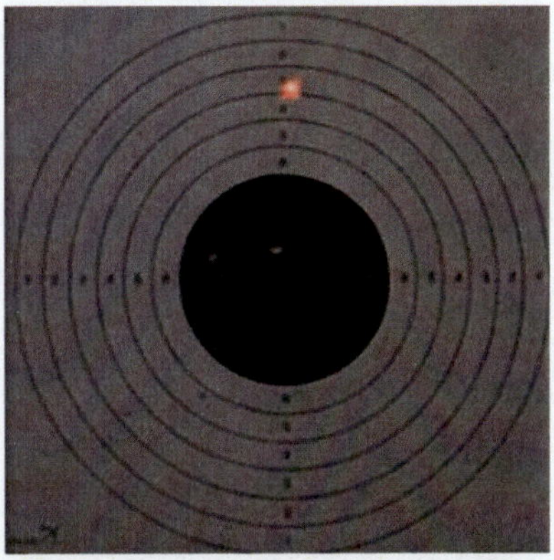

Fig. 4 Extracted target image

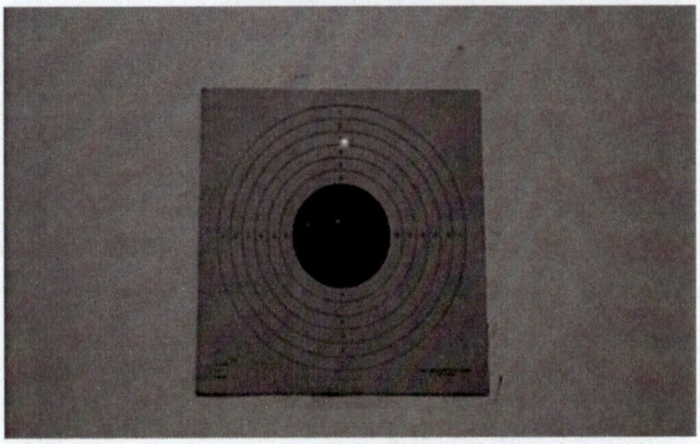

Fig. 5 Resized gray scale image

Fig. 6 Center of the target detected

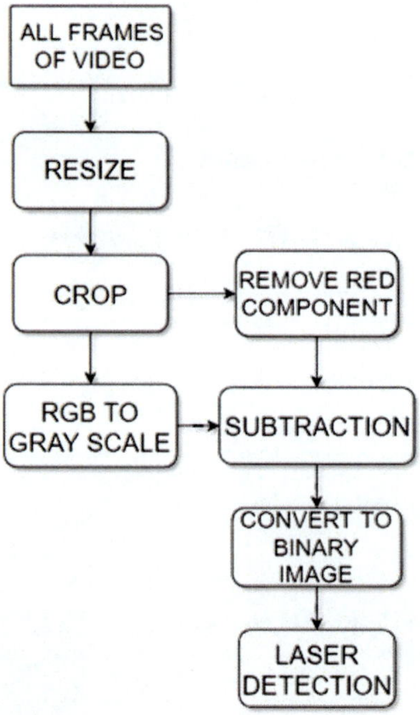

Fig. 7 Flow to detect the laser

Fig. 9. The reason for binarizing the image is to detect the laser. Once the laser gets detected, its center is the track point. The most important point is to consider the time delay between the trigger pressed and the shot detected. Thus, to minimize this error, the first frame has been taken into consideration. Here the first frame refers to the frame where we detect more than 1 center. The most critical concern is when more than two centers are detected (Fig. 10) or no center is detected, and to distinguish between the shot and tracking.

After analyzing a lot of data, the radius range and sensitivity are set according to the result of analysis of data, which clear the concern for no circle detection. It

Fig. 8 Detection of red spot

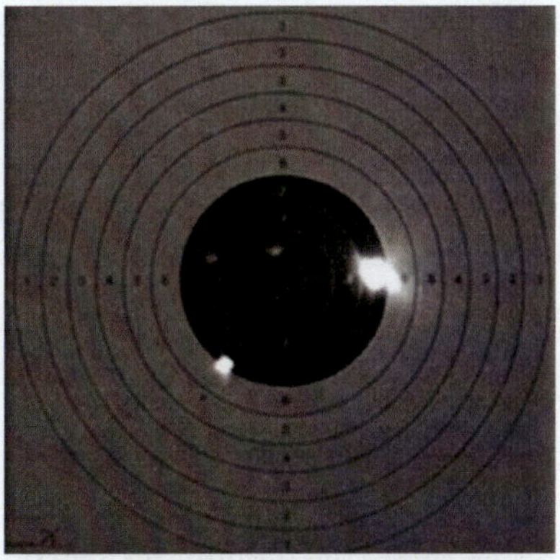

Fig. 9 Exact position of red laser

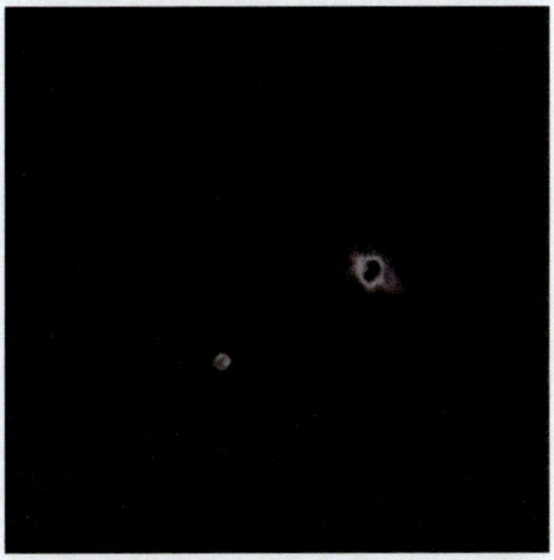

Fig. 10 Tracked laser

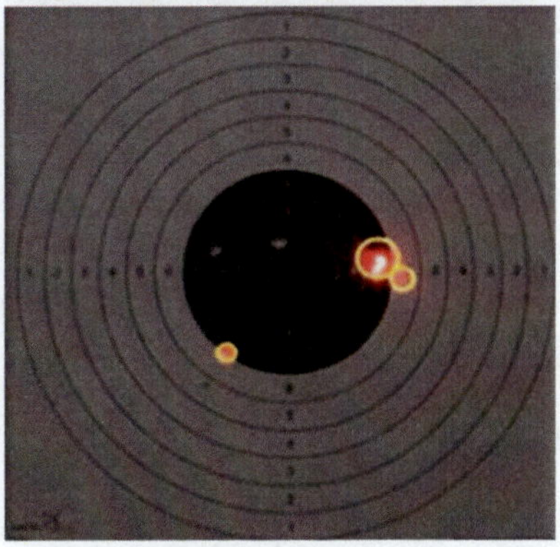

has been observed from the analysis that the circle with the least x coordinate is the actual spot of the laser. This issue was only encountered in the detection of the second laser due to malformation in the laser to make a proper circle.

For decoding the situation between shot and track, the constraint has been designed that the second laser should always point the right of the first laser, i.e., whenever the second laser gets triggered it should be in the right of the principal laser.

The algorithm for distinguishing the shot and the laser:

1. Unmark the shot
2. Check whether the circle is detected or not
3. If the circle is detected
    - 3.1 Find the minimum value of the center along x coordinate
    - 3.2 Find how many circles are detected
    - 3.3 If more than one circles are detected and the shot is unmarked
        - 3.3.1 Capture the frame
        - 3.3.2 Mark the shot.
    - 3.4 Otherwise unmark the shot
4 Note the x and y coordinate.

3.3 Analysis of Shot

The idea here is to alarm shooter about their behavior during the shot. By using the concept of traffic light signal, the shooter gets the indication of the shot. The shooter is also provided with the information of how much time they are spending in each area as shown in Figs. 11 and 12.

Indicator

- Red-> Start
  The red color is indicated whenever the shot is in the white area of the target.
- Yellow-> Ready
  The yellow color is indicated as soon as shooter enters the black area of the target.
- Green-> Fire
  The green color is indicated as shooter enters the aiming area.
- Maroon-> Cancel
  The maroon color is indicated once the shooter goes back to the black area above the center of the target after coming from the aiming area.

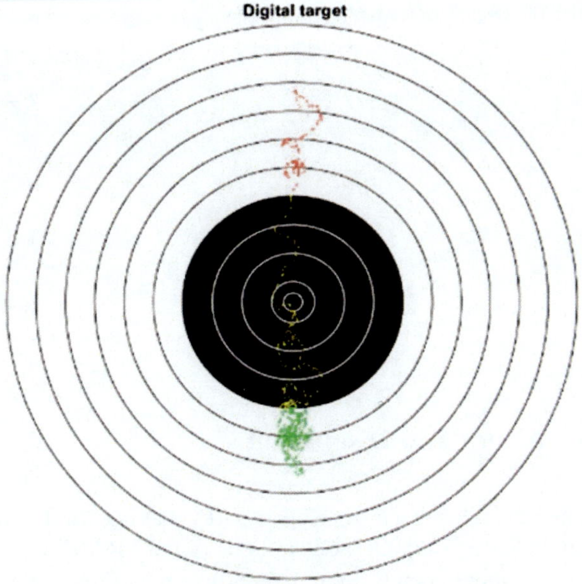

Fig. 11 Movement analysis of the gun

Fig. 12 Summary of time spend in each area of the target

	Position	Time	Indication
1	RED	1.6600	Start
2	YELLOW	4.3400	Ready
3	GREEN	5.6000	Fire
4	MAROON	0	Cancel

3.4 Construction of Digital Target

A white image as the background is being taken to form the target. The most important point to consider while forming the target is to match the center of the target with respect to the original target and conversion of millimeters (mm) to the pixel.

- Calculate the pixel per mm by formula

$$\text{Pixel Per MM} = r/59.5 \tag{1}$$

r = radius detected in finding the target.
59.5–(the radius of the 7th ring according to ISSF rule book).

- Calculate the radius of each ring from the ISSF rule book
- Draw the circle for each ring
- Draw the filled circle for the 7th ring (Fig. 13).

Fig. 13 Digital target

4 Result and Discussion

The model has been developed on PC having MATLAB version 9.5(R2018b) with Intel CORE i7(6500U) CPU, 8 GB (DDR3 RAM) and NVIDIA GTX 950 M GPU.

The shot depends upon the aiming area of the shooter which is between (4th and 6th ring). The system asks the shooter for the aiming area and depending on it the shot gets detected. The aiming area acts as a reference to detect the actual position of the shot, i.e., if the shooter chooses the aiming area as 5th ring then we need to shift the shot coordinate by the radius of the 5th ring according to ISSF rule book. In

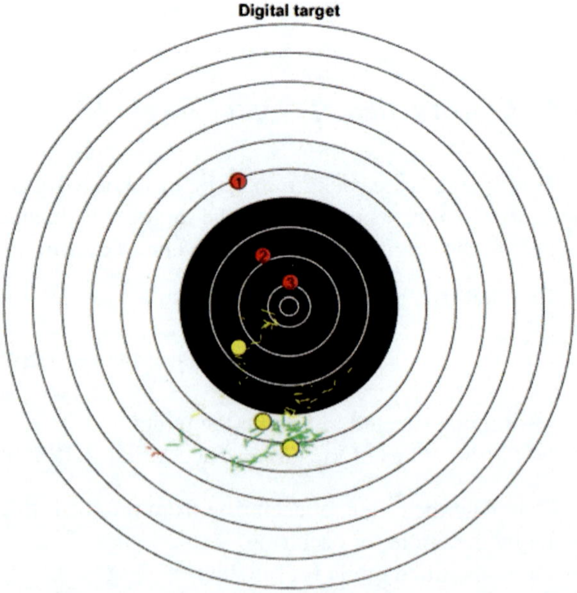

Fig. 14 Aiming area, shot fired and movement of the gun

Fig. 14, the red spot shows the shifted shot and yellow spot shows the detected shot. In Fig. 14, aiming area is shown as 5th ring.

Formation of Path.

All the position of a laser is known. The line is constructed from the coordinate obtained in section and changes in the color of the line are made according to details mention in Sect. 3.3.

Plot of the shot

- Center of the shot = Coordinate detected + radius of the aiming area(along y-axis only)
- Radius of a shot = radius of the bullet given in ISSF rule book.
- Plot the filled circle of the shot.

Calculation of shot

The shot score has been calculated up to one decimal value as per the rule is given by ISSF. The shot can fluctuate from 10.9 (Maximum) to 0.0 (Minimum). To get the accuracy up to one decimal, the distance between the two consecutive rings is split into nine parts. An important detail to mark is that the distance within the inner and outer 10 rings is not the same. Thus, to calculate the score between these rings will be different.

Formula

$$\text{Shot} = \begin{cases} 10 - \frac{(d-r)-r_{10}}{d2} D < r_{10} \\ 10 + \frac{9*\left(1-\frac{d}{r_{10}}\right)}{10} D \geq r_{10} \end{cases} \quad \ldots \tag{2}$$

d = Distance between center of target and pellet center.
r = Radius of the bullet.
r_{10}= Radius of the outer 10 ring.
d2 = Difference between two rings.

The above formula is being derived and 100% accuracy is been observed while testing on the various value of the shot and the value is been converted to one decimal place as per ISSF rule.

In Fig. 14, yellow filled circle is made from coordinate detected in Sect. 3.2. The red filled circle is made after considering the aiming area of the shooter. Here the aiming area considered is the 5th ring (Fig. 15).

5 Conclusion and Future Work

The shooting training system can improve the performance of the shooter. The system is very cost-effective and environment-friendly as it only requires a laser and the camera. Further, no more bullets are required for practice. Detection of the shot

Fig. 15 Analysis of the shots

	Position	Time	Indication
1	RED	0.1001	Start
2	YELLOW	1.6011	Ready
3	GREEN	3.0688	Fire
4	MAROON	0	Cancel

	Shot	Score
1	1	6.4000
2	2	9
3	3	10.2000

and tracking the movement of the gun is highly accurate, which make this model more reliable. Analysis of the model provides the shooter depth knowledge about the behavior of the gun and the movement of the body.

The model can be made smarter by applying the deep learning concept to train the model on the data set based on the activity in the brain and the movement in the body which will drastically improve the performance.

The model can be more advanced by changing the red laser to an infrared laser beam. A frequency toner can be used to detect the shot by taking the sound of the trigger into account, which will improve accuracy and time delay for the detection of the shot.

References

1. Ding P, Zhang X, Fan X, Cheng Q (2009) Design of automatic target-scoring system of shooting game based on computer vision. In: IEEE International Conference on Automation and Logistics, 2009, ICAL'09, pp 825–830. IEEE
2. Soetedjo A, Nurcahyo E (2011) Developing of low-cost vision-based shooting range simulator. Int J Comput Sci Network Secur 11(2):109–113
3. Guthe SA, Soni PM (2016) Target shooting training and instructive system model using python. In: International Journal of Engineering Research and Technology, pp 594–597
4. Rudzinski J, Luckner M (2012) Automatic scoring of shooting targets with tournament precision. In: KES, pp 324–334
5. Duda RO, Hart PE (1971) Use of the hough transformation to detect lines and curves in pictures. No. SRI-TN-36. Sri International Menlo Park CA Artificial Intelligence Center.
6. International Shooting Sport Federation, Official Statutes Rules and Regulation Sedition (2017)
7. Rongxin Du, Intelligent target scoring system based on image processing.
8. SCATT Company, Shooting training system user manual (SCATT MX–02)

9. Richard Pitcher, T (2015) An analysis of computer vision techniques to electronically score a rifle target. Rhodes University
10. SCATT Shooting Trainers. https://www.scatt.com/. Last accessed 20 Dec 2018

Integration of Data Mining Classification Techniques and Ensemble Learning for Predicting the Type of Breast Cancer Recurrence

Luis Luque, Reynaldo Villareal–González, and Luis Cabás Vásquez

Abstract The Wisconsin Breast Cancer Epidemiologic Simulation Model (Tsai et al. in J Big Data 2:21, 2015) is a stochastic simulation model that uses a scientific modeling system to study the level of breast cancer incidence and mortality of the population in the USA between 1975 and 2000. Four interactive processes are modeled simultaneously: Natural History of Breast Cancer, Breast Cancer Detection, Breast Cancer Treatment, and Breast Cancer Mortality. These components form a complex interactive system that simulates the lives of 2,950,000 women (approximately 1/50 of the U.S. population) from 1950 to 2000 in 6-month cycles. After a learning period of 25 years, the outputs of the model provide the incidence and mortality rates as a function of age between 1975 and 2000 (Guller in Big data analytics with spark: a practitioner's guide to using spark for large scale data analysis. Apress, Berzkeley, CA, USA, 2015). The model also simulates cases of disease, both detected and hidden, at the individual level, and can also be Used to answer questions about treatment selection and effectiveness, as well as to estimate the benefits for women of specific ages and histories.

Keywords Breast cancer detection · Breast cancer survival · Data mining · Weka · Learning

---

L. Luque (✉)
Universidad de La Costa (CUC), Barranquilla, Colombia
e-mail: lluque4@cuc.edu.co

R. Villareal–González
Universidad Simón Bolívar, Barranquilla, Colombia
e-mail: rvillareal2@unisimonbolivar.edu.co

L. C. Vásquez
Corporación Universitaria Latinoamericana, Barranquilla, Colombia
e-mail: lcabas@ul.edu.co

R. Patgiri et al. (eds.), *Proceedings of International Conference on Big Data, Machine Learning and Applications*, Lecture Notes in Networks and Systems 180,
https://doi.org/10.1007/978-981-33-4788-5_5

1 Introduction

In the last decade, Data Mining and Statistical Analysis has been widely used in the health care industry. When these methods are used in conjunction with information extracted from large amounts of data, they can help physicians make decisions and improve service.

Breast cancer affects approximately 15% of women with risk age worldwide [1]. Early detection of the presence of malignant cancer cells increases the chance of life for patients, especially when a small, unbranched tumor is located.

The objective of this paper is to analyze the data corresponding to a group of women in order to obtain certain patterns when applying them to detect possible cases of breast cancer through data mining, similar to other studies based on genetic algorithms [2, 3].

2 Data Mining

Data mining (DM) is the nontrivial extraction of information that is implicit in data. Such information was previously unknown and may be useful for a process. In other words, data mining prepares, tests, and explores data to extract the information hidden in it [4]. See Fig. 1.

3 Database

The Wisconsin Breast Cancer Database consists of 699 cases, each with 9 attributes corresponding to subjective observations of the tumors, plus two attributes corresponding to case identification and classification of the tumor (benign or malignant) [6]. The subjective observations are based on descriptions of the cells obtained from an image.

This approach follows a seven-step process to extract useful information from the data [7].

1. Identify the targets
2. Obtain a dataset to be analyzed
3. Data pre-processing
4. Data Transformation
5. Data Mining and Statistical Analysis
6. Interpretation and evaluation
7. Writing the report.

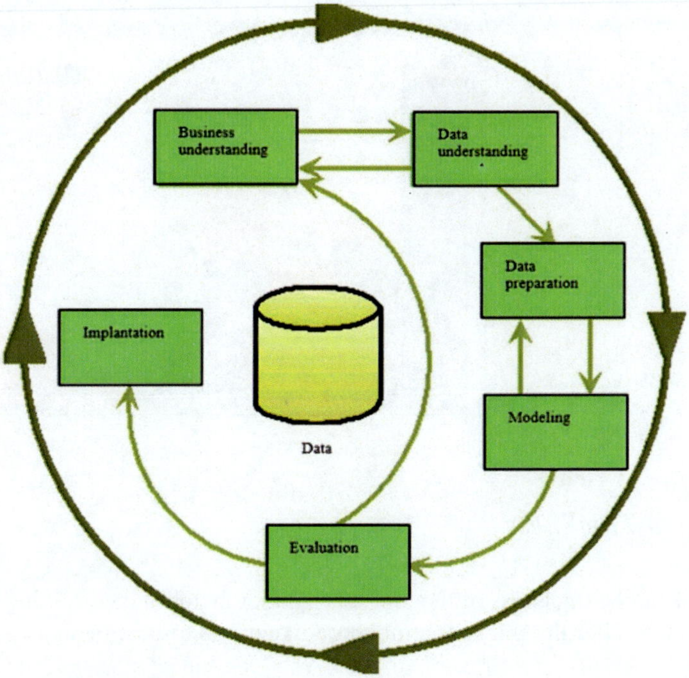

Fig. 1 Data mining project life cycle [5]

3.1 Attributes

1. Sample code number: identification code of each patient [4].
2. Clump thickness: thickness of clump.
3. Uniformity of cell size, size of cell.
4. Uniformity of cell shape: shape of cell.
5. Marginal adhesion: marginal adhesion.
6. Single epithelial cell size: individual cell size [8].
7. Bar nuclei: nucleus.
8. Bland chromatin: soft chromatin.
9. Normal nucleoli: normal nucleoli.
10. Mitoses: mitosis.
11. Class: {benign, malignant}.

The class distribution is: benign 458 (65.5%) and malignant 241 (34.5%).

Fig. 2 Thickness

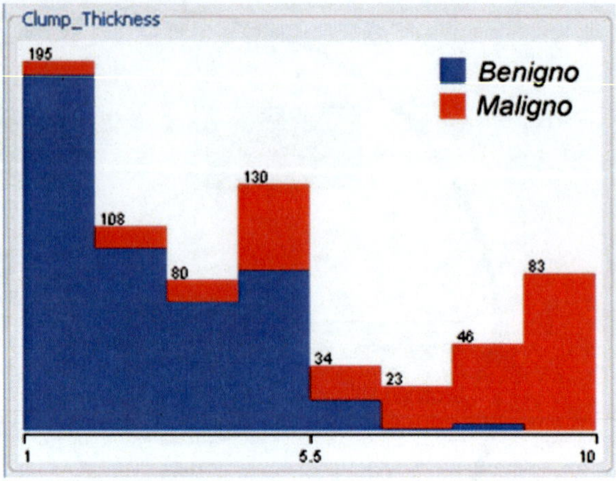

3.2 Clump Thickness

As can be seen, the thickness of the breast is a major factor in determining the type of tumor; the smaller the thickness, the more likely it is to be a benign cancer [9]. See Fig. 2.

3.3 Uniformity of Cell Size

Benign cases are concentrated in the uniformity index 1. See Fig. 3.

Fig. 3 Uniformity of cell size

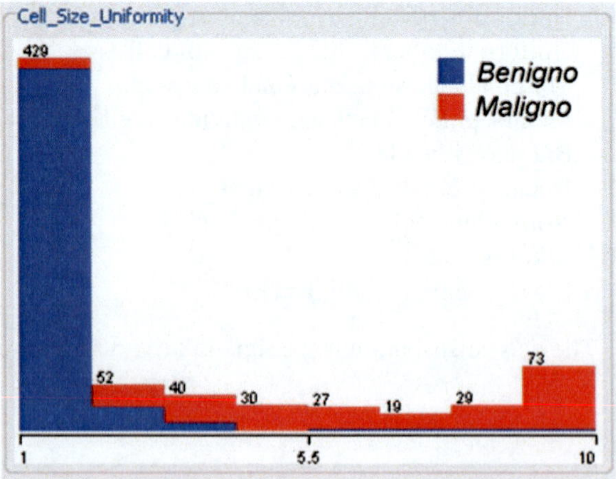

Fig. 4 Individual cell size

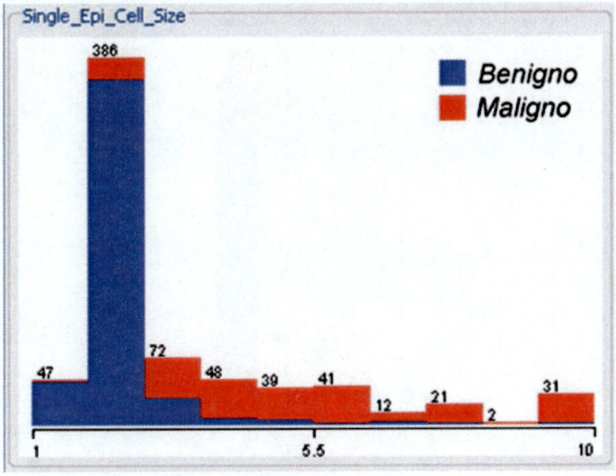

3.4 *Single Epithelial Cell Size*

In this case, most cell sizes are concentrated in small values, with the smallest size being related to whether the tumor is benign [10]. See Fig. 4.

3.5 *Other Considerations*

The rest of the attributes presents similar histograms (the lower value plus the number of benign ones), although it should be noted that the Mitoses attribute [11] is apparently not informative because the vast majority of cases (both benign and malignant) are concentrated at the lowest level. See Fig. 5.

4 Evaluation of the Data

This would already be the fourth of the steps to be taken when carrying out a study on data mining. For this reason, the results sought for this study will be extracted at each moment according to the parameters. For this purpose, the different classification tools offered by Weka will be used [12].

4.1 *K-NN*

In the K-NN method [13] (K-Nearest neighbors), it is a supervised sorting method.

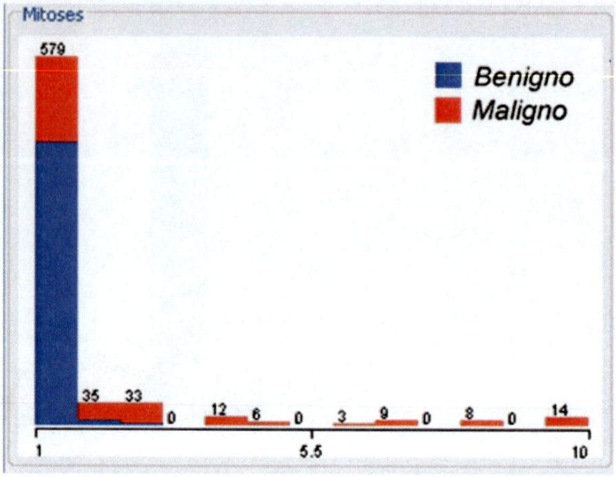

Fig. 5 Mitosis

This is a nonparametric classification method, which estimates the value of the probability density function or directly the ex-post probability that an x-element belongs to a certain Cj-class from the information provided by the sample set. In general, the Euclidean distance will be used to determine the class [14].

The optimal k value that should be set depends on the data, since a too small value would make the system very sensitive to noise and a too high value would lead to a bad classification.

In the previous example, one can see how a new sample (in green) is to be classified with a k-NN. It can be seen that if a $k = 3$ is chosen, the new sample will be classified in the red class, whereas if a $k = 5$ is chosen, the sample will be classified in the blue class.

In this case, the IB-k function available in Weka will be used, with a k within the interval [1, 10], activating the cross-validation box with 10 folds, obtaining the following results [15].

According to the previous graph, the error fluctuates according to the chosen k. In this case, the optimal k is produced for $k = 3$, producing an error of 3.21%. See Fig. 6.

4.2 *MlP*

The multilayer perceptron [16] is an artificial neural network (ANN) formed by multiple layers, allowing it to solve problems that are not linearly separable, which is the main limitation of the perceptron. Multilayer perceptron can be totally or locally connected. In the first case, each output of an "i" layer neuron is input from all "i+1" layer neurons, while in the second case, each "i" layer neuron is input from a series of "i+1" layer neurons (region) [17]. See Fig. 7.

The layers can be classified into three types [18].

Fig. 6 K-NN classification

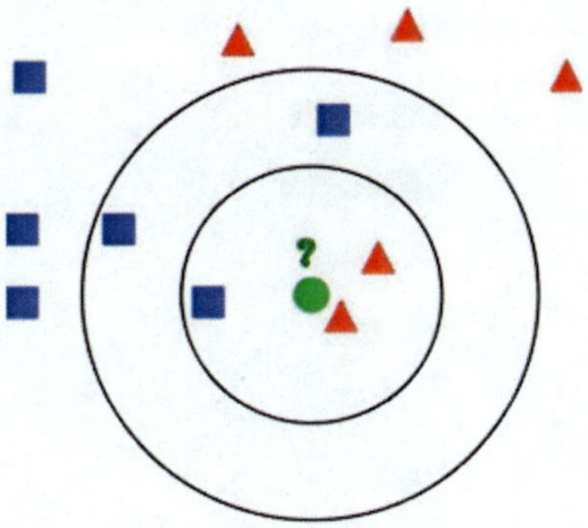

Fig. 7 K-NN error

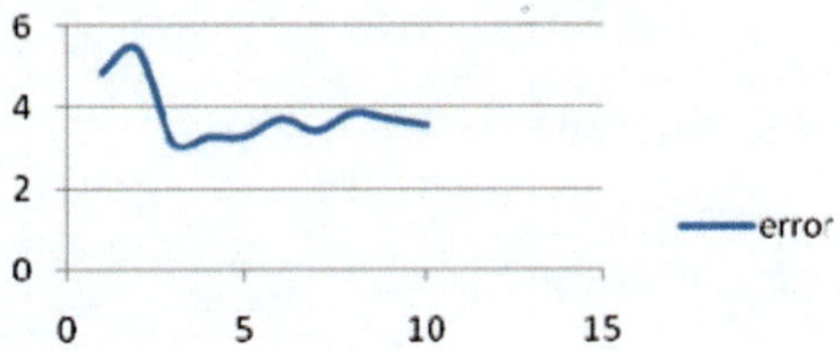

1. Input layer: Made up of those neurons that introduce the input patterns into the network. No processing takes place in these neurons.
2. Hidden layers: Formed by those neurons whose inputs come from previous layers and the outputs go to neurons in later layers.
3. Output layer: Neurons whose output values correspond to the outputs of the entire network.

The neural network layers used in this application are of two types “a”, “t” and “i”, where ‘a’ = (attribs + classes) / 2, ‘i’ = attribs, ‘o’ = classes, ‘t’ = attribs + classes. Class a can be seen in Fig. 8.

The results obtained for each different implementation are as follows (Tables 1, 2 and 3).

Class a MLP. See Table 1.

Where an error of 4,145% is obtained.

Class i MLP. See Table 2.

Where an error of 4,10% is obtained. See Table 3.

Class t MLP.

Where an error of 4.99% is obtained.

As can be seen, the three values are quite similar, the failure rate is around 4–5%, the class I perceptron being better. The number of hidden layers of the MLP could be increased even more in order to have a much more reliable and closer to real model, although this would lead to an increase in computation time and complexity.

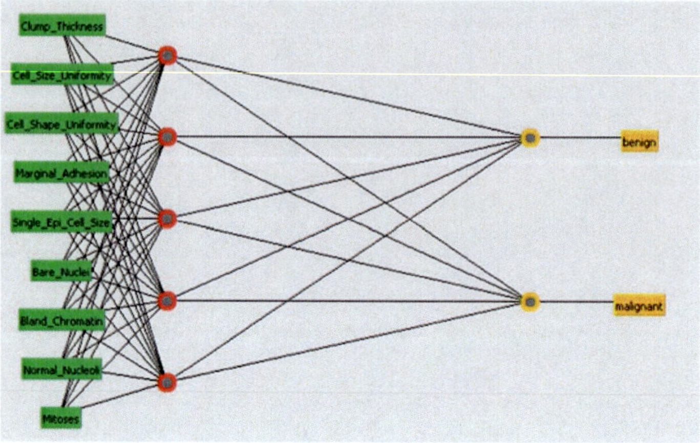

Fig. 8 Class a MLP [18]

Table 1 Class a MLP results

Benign	Malignant	<-Classified as
521	21	Benign
18	321	Malignant

Table 2 Class i MLP Results

Benign	Malignant	<-Classified as
524	19	Benign
15	310	Malignant

Table 3 Class t MLP results

Benign	Malignant	<-Classified as
527	20	Benign
19	325	Malignant

4.3 Naïve Bayes

Naïve Bayes (NB) [19, 20] is a probabilistic classifier that assumes that each parameter–value pair in the same example is independent from the rest. Each parameter–value pair is assigned a probability of belonging to a class. This is done by dividing the number of examples in each class in which this pair appears by the number of examples that belong to that class. In order to classify a new case, the probability of belonging to each class is calculated, classifying it in the class where the probability is greater, thus adopting a maximum a posteriori estimation criterion.

This probability of membership is calculated as the product of the probability of membership in each class for each of the parameter–value pairs that define the case

Table 4 Naïve bayes results

Benign	Malignant	<-Classified as
521	20	Benign
16	310	Malignant

Table 5 SMO results

Benign	Malignant	<-Classified as
519	22	Benign
18	315	Malignant

Table 6 Results j48

Benign	Malignant	< -Classified as
530	24	Benign
21	320	Malignant

to be classified. And it would be classified according to the highest value above. See Table 4.

By taking it to Weka and using the Simple Naïve Bayes function, it is obtained that:

Correctly Classified Instances 752 95.99852
Incorrectly Classified Instances 35 4.00148.

As can be seen, the classification error for this method is similar to that obtained for the class I perceptron, although it can be observed that the failure rate is distributed differently, with the false alarm rate now being much higher than the loss rate, with only 6 cases being wrongly classified.

4.4 SMO

SMO implements the minimum sequential optimization algorithm to train a Vector Support Machine (VSM) [14]. These algorithms extend the characteristics of linear models, since they allow distinguishing between classes that present nonlinear decision limits. For this purpose, the original data are transformed in a nonlinear way into a new, larger space. In this new space, a linear model is built to represent a nonlinear decision boundary in the original space. A more detailed introduction to SVM can be found in [18] (Table 5).

Correctly Classified Instances 754 97.3241%
Incorrectly Classified Instances 33 2.6759%.

As can be seen in the results obtained by SMO, the failure rate is reduced by up to 3.14%, even below 3-NN. On the other hand, having a lower error rate, the probability of loss is higher than in the previous case, which is not advantageous. It should also be noted that the probability of false alarm is reduced, to the point that there are only 14 cases that contemplate it.

4.5 RFB

RFB [14, 20] is an alternative method to multilayer perceptron (MLP) (e.g., backpropagation) for adjusting to non-linear functions. Unlike the arrangement of activation functions that allows building training models by backpropagation, these new RBF-based networks build their models with activation functions that are different in both the hidden and the output layer. That is, an RBF network is designed with neurons in the hidden layer activated by nonlinear radial functions with their own gravitational centers and in the output layer by linear functions.

In Weka, it uses a k-means clustering algorithm and learns by logistic or linear regression. It will use the number of clusters of the classes to be classified, and will standardize the data to mean zero and unit variance. Performing in the same way as in K-NN, the following failure rates are obtained as a function of the number of clusters (Fig. 9).

Where one can see that the error increases if the number of clusters is increased. For the minimum error, the following result is available (Table 6).

Correctly Classified Instances 670 95.8512%
Incorrectly Classified Instances 29 4.1488%.

As can be seen, the RFB network gives a good result when detecting positive cases, where it can be seen that the probability of losses is small, although the probability of false alarms is high compared with the other analyzed algorithms.

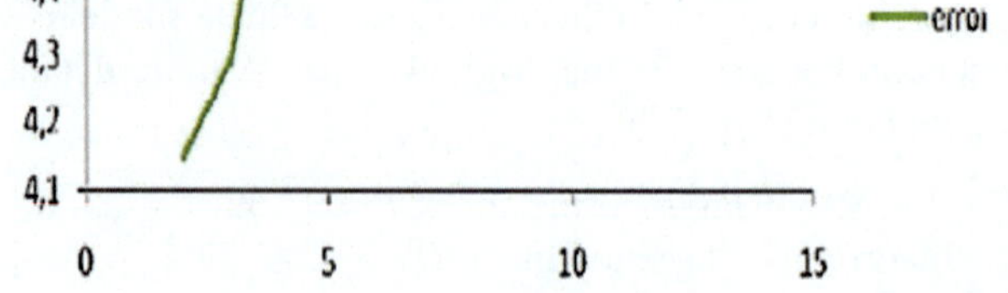

Fig. 9 RFB error (number of cluster)

4.6 J48

J48 [12] implements the public version of the classification algorithm based on decision trees C4.5 revision 8, prior to the release of version C5.0 [21]. These trees classify a new case by evaluating, in each node of the model, the parameters that define the case to be classified. The cases that, starting from the root, reach a certain leaf receive the classification indicated by the leaf.

The J48 algorithm is based on the use of the gain ratio criterion. This prevents the variables with the largest number of possible values from being selected. In addition, the algorithm incorporates a pruning of the classification tree once it is induced. Executing the algorithm in Weka, it is obtained:

Correctly Classified Instances 760 94.5478%
Incorrectly Classified Instances 35 5.4522%.

As can be seen, with this method, the error rate rises to 5.35%.

5 Conclusions

In view of the results obtained, it was concluded that although all the attributes can affect to a greater or lesser extent the classification of a tumor as benign or malignant, and that, in general, the lower the value of the attribute, the greater the probability that it is a benign class. In the specific case of the Mitosis attribute, as was mentioned at the time, it can be thought that its elimination from the study will not greatly affect the results, since most cases, both of one class and another are at the lowest level.

Analyzing the results of the different classifiers, slight differences can be seen between them, although the error rates of all of them are concentrated around 3–5%, in a more detailed analysis.

- K-NN has the advantage of allowing to choose the K value according to the requirements, however, there is a problem of finding the optimum value and it is necessary to vary it.
- The multilayer perceptron classifier allows to select among several options, although the results among them ares similar and present worse characteristics than, for example, the K-NN.
- Naïve Bayes allows to obtain a higher probability of error than for most K-NN, although the distribution of errors tends to classify malignant tumors as benign, which could be considered more serious than the opposite case. The SMO classifier has the best error rate, with errors fairly balanced between false alarms and losses.
- The minimal error case of the RFB is similar to that of the Naïve Bayes, with a much higher number of false alarms than losses, which is a problem. For higher values of the number of clusters, the error rate increases.

- Considering the above, the use of the SMO classifier would be preferred, due to its low error rate and the balance between false alarms and losses.

References

1. Tsai CW, Lai CF, Chao HC, Vasilakos AV (2015) Big data analytics: a survey. J Big Data 2(1):21
2. Guller M (2015) Big data analytics with spark: a practitioner's guide to using spark for large scale data analysis. Apress, Berkeley, CA USA
3. Shah C, Jivani AG (2013) Comparison of data mining classification algorithms for breast cancer prediction. In: 2013 Fourth international conference on computing, communications and networking technologies (ICCCNT), pp 1–4. IEEE
4. Viloria A, Viviana Robayo P (2016). Virtual network level of application composed IP networks connected with systems—(NETS Peer-to- Peer). Indian J Sci Technol 9(46). https://doi.org/10.17485/ijst/2016/v9i46/107376
5. International Agency for Research on Cancer (IARC) and World Health Organization (WHO) (2018) Globocan 2018: Age Standardized (World) Incidence and Mortality Rates Breast, Sep. 2018, [online] Available https://gco.iarc.fr/today/data/factsheets/cancers/20-Breast-fact-sheet.pdf
6. Sakri SB, Rashid NBA, Zain ZM (2018) Particle swarm optimization feature selection for breast cancer recurrence prediction. IEEE Access 6:29637–29647
7. Moreno A, Redondo T (2016) Text Analytics: the convergence of Big Data and Artificial Intelligence. In: At International Journal of Interactive Multimedia and Artificial Intelligence
8. Shukla N, Hagenbuchner M, Win KT, Yang J (2018) Breast cancer data analysis for survivability studies and prediction. Comput Methods Programs Biomed 155:199–208
9. Silva J et al (2019) Psychosocial analysis of policy confidence through multifactorial statistics. In: At International Conference on Computational Vision and Bio Inspired Computing, pp 1–7. Springer, Cham
10. Zheng B, Yoon SW, Lam SS (2014) Breast cancer diagnosis based on feature extraction using a hybrid of K-means and support vector machine algorithms. Expert Syst Appl 41:1476–1482
11. Mohebian MR, Marateb HR, Mansourian M, Mañanas MA, Mokarian F (2017) A hybrid computer-aided-diagnosis system for prediction of breast cancer recurrence (HPBCR) using optimized ensemble learning. Comput Struct Biotechnol J 15:75–85
12. Lu Y, Han J (2003) Cancer classification using gene expression data. Inf Syst 28:243–268
13. Witten IH, Frank E (2005) Data mining: practical machine learning tools and techniques, 2nd edn. Morgan Kaufmann series in data Management Systems. ISBN 0-12-088407-0
14. Yue W (2016) Research on the clustering analysis algorithm for data mining. Revista Ibérica De Sistemas Y Tecnologías De La Información (RISTI) E6:209–221
15. Li D, Park HW, Batbaatar E, Piao Y, Ryu KH (2016) Design of health care system for disease detection and prediction on Hadoop using DM techniques. In: Conference on health informatics and medical systems
16. Apache Spark—Unified Analytics Engine for Big Dat, (2018) Available https://spark.apache.org/
17. Ray RB, Kumar M, Rath SK (2016) Fast in-memory cluster computing of sizeable microarray using spark. In: Proceeding International Conference Recent Trends Information Technology (ICRTIT), pp 1–6
18. Silva J, Varela N, Lezama OBP, Hernández-P H, Ventura JM, de la Hoz B, Coronel LP (2019) Multi-dimension tensor factorization collaborative filtering recommendation for academic profiles. In: International Symposium on Neural Networks, pp 200–209. Springer, Cham

19. Silva J, Solano D, Fernández C, Romero L, Lezama OBP (2019) Psychosocial analysis of policy confidence through multifactorial statistics. In: International conference on computational vision and bio inspired computing, pp 1–7. Springer, Cham
20. Viloria A, Lezama OBP, Varela N (2019) Method for estimating height in people using multivariable statistics. Procedia Comput Sci 160:224–228
21. Taneja A, Arora A (2018) Cross domain recommendation using multidimensional tensor factorization. Expert Syst Appl 92(1):304–316

Image Reconstruction Based on Shape Analysis

Shalini Ramanathan and Mohan Ramasundaram

Abstract In recent years, image reconstruction based on graph theory has aroused growing interest. Here a novel MRS framework was introduced. It works to reconstruct the brain image based on shape analysis. The shape was recognized from the input image. Its description and count were collected in a hypergraph structure for the reconstruction process. The MRS framework reconstructs an image based on the identified shapes. The reconstructed images help to locate the missing data and to identify the disorder in the brain.

Keywords Image reconstruction · Brain image · Hypergraph structure · Shape analysis · Brain disorders

1 Introduction

Biomedical images usually contain complex objects, which will vary in appearance significantly from one image to another. Attempting to measure or detect the presence of a particular structure in such images was a difficult task. However, by using models that can cope with the inconsistency, it is possible to analyze complex images successfully. Complex images give rise to many primitives, which must be linked together correctly to provide a successful interpretation of quality images [1]. Based on the World Alzheimer's Report, dementia was not a normal part of aging. So there is a need for "biomarkers," which are measurable indicators of a biological condition of the human brain [2, 3]. According to the India Aging Report 2019, more than 4

S. Ramanathan (✉) · M. Ramasundaram
Department of Computer Science and Engineering, National Institute of Technology Tiruchirappalli, Tiruchirappalli 620015, Tamil Nadu, India
e-mail: 406916001@nitt.edu

M. Ramasundaram
e-mail: rmohan@nitt.edu

R. Patgiri et al. (eds.), *Proceedings of International Conference on Big Data, Machine Learning and Applications*, Lecture Notes in Networks and Systems 180,
https://doi.org/10.1007/978-981-33-4788-5_6

million people had some form of dementia, which was a burden for old aged people. The inverse problem is the identification of the input of a given image from the knowledge of the output [4, 5]. The reconstruction of an image from the acquired data is an inverse problem. Image reconstruction is a critical problem in medical image analysis [6–9]. In this paper, a novel hypergraph based shape reconstruction framework for Medical Diagnosis named MRS framework was introduced. The low-level brain image data was examined to find shapes of euclidean geometry. Then those shapes were assembled into groups in an attempt to identify objects of interest. Image reconstruction helps to improve the noisy, corrupted, and limited tomographic data in the medical image. Brain scan images were collected by monitoring the patients at a regular interval. In each image, there was an identified damage. All the images were grouped together to find the number of injuries in the whole region of the brain.

2 Shape Reconstruction

2.1 Preamble

Initially, the MRS framework identifies the error-free shape. It was placed in the CSV file and provided as the next level input to the framework. All the identified shapes are collected in the edges of the hypergraph cluster [10–12]. Where each edge represents the shape name, for example, the line was denoted as $E1$ and the triangle was denoted as $E2$ and so on. Each vertex represents the shape number. For example, the collection $L1, L2, T1, T2, T3$ represents two lines and three triangles.

2.2 Data Collection and System

MGH data from the Human Connectome Project was obtained with the help of connectome Workbench [13]. Twenty-four subjects with 3T1 brain MRI scan image data used for assessment of image reconstruction. This image consists of 6 age groups of human lifespan in 3 volumes; each volume has 128 different slices of the human brain. The images are in NIfTI format and single-channel grayscale images. The model was trained on a system with Windows 10 of Intel Core i7 CPU 3.40 GHz and NVidia GeForce GT 730 processor with 6GB RAM. The model was implemented in python 3.7 in CUDA visible system environment [14]. All the 128 scan image of the human brain from MGH was placed in MRS Framework for preprocessing.

2.3 Data Preprocessing

Brain image data collection was preprocessed in order to produce a noise-free image. The basic shape point was considered an error; that is, it was the noise in the image. All images in the NIfTI format were converted to jpeg format. They were examined by the framework for the presence of point, and if present, then it was removed from the input collection. After preprocessing the images, the new image collections were fed into the next module, hypergraph cluster, to generate or reconstruct a shape [15].

2.4 MRS-Hypergraph Framework

The MRS-Hypergraph framework was executed for smiley shape reconstruction with the shape analysis method before testing with a real-time image. The real-time results of the framework were discussed in Sect. 3.2. In the example of smiley shape reconstruction, the notation "N" denotes the total count of an input image. The N numbers of input images were placed in CSV files; each image represented with N $\times$ M (here, it is 28 $\times$ 32) matrices. Each input file was read to locate the first cell with binary digit "1". The shape was detected, and the information about the name of the shape and its position were collected in a hypergraph structure. The stored structure was sorted according to the first position; also the first dimension of an array, to reconstruct an image with smiley shaped. Check the nodes of the graph for availability, whether there were exactly three horizontal lines and exactly one non-horizontal line or vertical line. Save this information in a tuple form. Now, the tuple was used to reconstruct the correct smiley image with the following error function.

Let $x_(i, j)$ represent the input matrix, where i = 1,...,m and j = 1,...,n. Assume $h_1 h_2 h_3$ are the three horizontal lines and v represents a vertical line.

$$f(s) = \begin{cases} h_1(x_i) \neq h_2(x_i) \\ h_1(x_i) \geq v(x_i) \\ h_3(x_i) \geq v(x_i) \\ h_1(x_j) > v(x_j) \forall j = 1, ..., m \\ (\exists v(x_j) < h_3(x_{p-1})) and (\exists v(x_j) < h_3(x_{p-n})) \forall j = 1, ..., m \end{cases} \quad (1)$$

The smiley error function was created in Eq. 1 to check, whether the reconstructed image was an exact shape of smiley or not. The Eq. 1) was generated based on the following conditions. That is, the first two horizontal lines should be on the same axis in the first half of the matrix so that it represents the "eye" of the smiley. And at the same time, it should be above the single vertical line. The third horizontal line should be below the vertical line and not start from the vertical lines' y-axis. The vertical line should be enclosed between the pair of the first two horizontal lines and

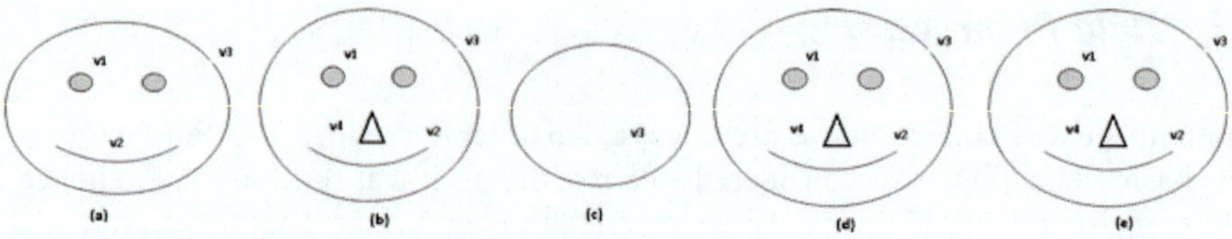

Fig. 1 **a–e** Input image generated with basic shapes

the third horizontal line. Endpoints of the vertical line should be between the pair of the first two horizontal lines. Also, the vertical line should be between the third horizontal lines. If the error function was false, for all the conditions given in Eq. 1), then it was a correct shape of a smiley.

3 Results and Inference

Initially, the framework was tested against the few self-created input images, it was discussed in Sect. 3.1, then it was tested against the real-time brain image and the results were conferred in Sect. 3.2.

3.1 Smiley Shape Reconstruction

The basic shapes like circle, line, point, and triangle were given as an input to generate a face image [17]. The generated face image was represented as an incidence matrix. In a similar way, five different 28 X 28 size matrix was created. Each image contains a smiley shape with missing data. This collection of smiley images was the input to the framework.

The MRS framework reconstructed a smiley shape, and it was shown in Fig. 1e. The reconstruction operation also removes the noise that removes the point shape, present in the image. Each image number was represented as a hypergraph edge, and shape components were represented as a hypergraph vertex. Figure 1a–d was the collection of input images with missing data. Figure 1e was the generated output image.

Table 1 shows the reconstructed final image based on hypergraph construction, vertices are collected in the same edge and the edges are joined together to form a hypergraph structure, which represents the reproduced image with the presence of all shapes.

Table 1 Hypergraph representation with hyper vertex and hyper edges

Input image number	Hper-vertex and Hyper-edges	Shapes used to create an image
1	E1=v1,v2,v3	Filled circle, Curve, Circle
2	E2=v1,v2,v3,v4	Filled circle, Curve, Circle, Triangle
3	E3=v3	Circle
4	E4=v1,v2,v3,v4	Filled circle, Curve, Circle, Triangle
Final	v1,v2,v3,v4	Filled circle,Curve, Circle, Triangle

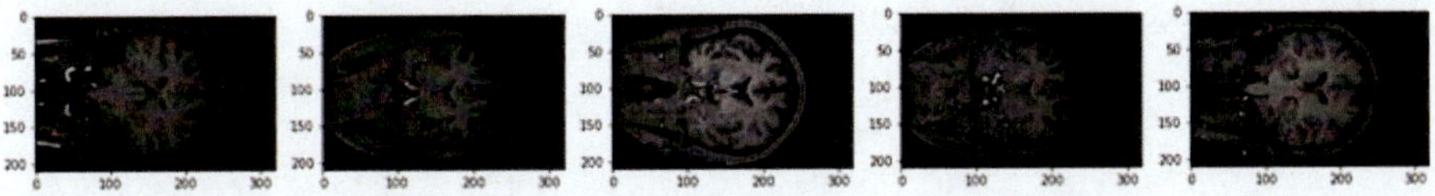

Fig. 2 Ground truth image of brain MRI scan from HCP-MGH

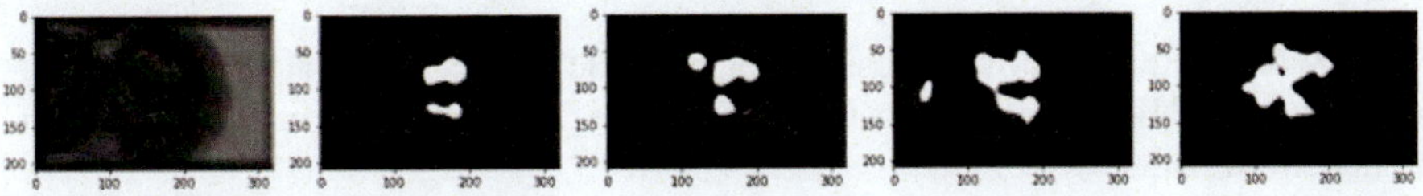

Fig. 3 Reconstructed image as output from MRS Framework with removed noise

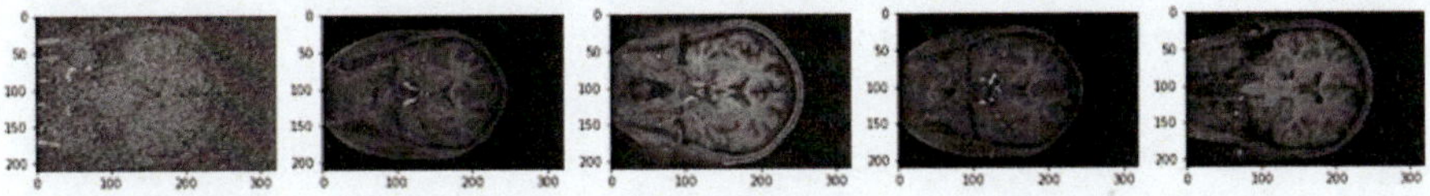

Fig. 4 Reconstructed image as output from MRS Framework without removing noise

3.2 Brain MRI Scan Image of HCP-MGH

It has been inferred from the smiley shape reconstruction, the reconstruction process of the MRS framework was able to find a pattern and also able to produce a new image from a collection of the input image. Now the MRS framework was tested against real-time images collected from the Human Connectome Project-Massachusetts General Hospital database [17, 18]. Ground truth image of brain MRI scan from HCP-MGH was shown in Fig. 2. The reconstructed images without removing noise and images with removed noise shown in Figs. 3 and 4.

Table 2 Evaluation metric done by MRS framework

Approaches	Proposed MRS approach	Single decoder	ScSR [20]	3D-CNN [21]
PSNR (in db)	39.7 dB	36.96 dB	35.96 dB	34.20 dB

3.3 Performance Measure

To evaluate the reconstructed image's best quality, the mean square error function named the PSNR- Peak Signal to Noise Ratio (in dB) calculated by the framework with the following formula given in Eqs. 2 and 3.

$$MSE = \left\{ \frac{1}{mn} \sum_{i=0}^{m-1} \sum_{j=0}^{n-1} [I(i, j) - K(i, j)]^2 \right. \quad (2)$$

$$PSNR = \left\{ 10 \; log_{10} \frac{MAX}{MSE} \right. \quad (3)$$

where MAX is the maximum possible pixel value of the image [19]. The typical value for the PSNR in the reconstructed image was 39.7 dB shown in Table 2, which infers better results.

3.4 Limitations and Applications

The MRS framework was developed for the targeted field of medical image processing. It should be enhanced further with unsupervised deep learning with autoencoders. It also has been enhanced based on the performance measure with an increase in PSNR value. In the field of unsupervised learning, by giving multiple images as input [20]. This framework will also be used to generate an image-based password.

4 Conclusion

Brain image was obtained from the Human Connectome Project database. Image reconstruction was applied to remove noise and reproduce the image through the MRS framework. Identified that the framework outperforms in finding the shape and number of shapes present in the input image. The framework also well performed in reproducing the image for the real-time brain MRI image.

References

1. Krupinski EA (2010) Current perspectives in medical image perception NIH Public Access. Atten Percept Psychophys 72. https://doi.org/10.3758/APP.72.5.1205
2. Christina Patterson (2018) World Alzheimer Report 2018—The state of the art of dementia research: New frontiers; World Alzheimer Report 2018—The state of the art of dementia research: New frontiers
3. Alzheimer's Association (2019) 2019 Alzheimer's disease facts and figures includes a special report on Alzheimer's detection in the primary care setting: connecting patients and physicians
4. Beskos A, Girolami M, Lan S, Farrell PE, Stuart AM (2017) Geometric MCMC for infinite-dimensional inverse problems. J Comput Phys 335:327–351. https://doi.org/10.1016/J.JCP.2016.12.041
5. Gong K, Kim K, Cui J, Guo N, Catana C, Qi J, Li Q (2018) Learning personalized representation for inverse problems in medical imaging using deep neural network
6. Shen G, Horikawa T, Majima K, Kamitani Y (2019) Deep image reconstruction from human brain activity. PLOS Comput Biol 15:e1006633. https://doi.org/10.1371/journal.pcbi.1006633
7. Wang G (2016) A perspective on deep imaging. IEEE Access 4:8914–8924. https://doi.org/10.1109/ACCESS.2016.2624938
8. Wang G, Ye JC, Mueller K, Fessler JA (2018) Image reconstruction is a new frontier of machine learning. IEEE Trans Med Imaging 37:1289–1296. https://doi.org/10.1109/TMI.2018.2833635
9. Després P, Jia X (2017) A review of GPU-based medical image reconstruction. Phys Medica 42:76–92. https://doi.org/10.1016/J.EJMP.2017.07.024
10. Berge C (1989) hypergraphs: combinatorics of finite sets. North Holland
11. Somu N, Raman MRG, Kirthivasan K, Sriram VSS (2016) Hypergraph based feature selection technique for medical diagnosis. J Med Syst 40:239. https://doi.org/10.1007/s10916-016-0600-8
12. Zhou D, Huang J, Schölkopf B (2007) Learning with hypergraphs: clustering, classification, and embedding. In: 2006 Conference on Advances in Neural Information Processing Systems, pp 1601–1608. MIT Press
13. Van Essen DC, Ugurbil K, Auerbach E, Barch D, Behrens TEJ, Bucholz R, Chang A, Chen L, Corbetta M, Curtiss SW, Della Penna S, Feinberg D, Glasser MF, Harel N, Heath AC, Larson-Prior L, Marcus D, Michalareas G, Moeller S, Oostenveld R, Petersen SE, Prior F. Schlaggar BL, Smith SM, Snyder AZ, Xu J, Yacoub E, WU-Minn HCP Consortium W-MH (2012) The human connectome project: a data acquisition perspective. Neuroimage 62:2222–2231. https://doi.org/10.1016/j.neuroimage.2012.02.018
14. Mark Harris: Numba: High-Performance Python with CUDA Acceleration,NVIDIA Developer Blog. https://devblogs.nvidia.com/numba-python-cuda-acceleration
15. Bretto A, Cherifi H, Aboutajdine D (2002) Hypergraph imaging: an overview. Pattern Recognit 35:651–658. https://doi.org/10.1016/S0031-3203(01)00067-X
16. Zeng G (2001) Image reconstruction-a tutorial. Comput Med Imaging Graph 25:97–103. https://doi.org/10.1016/S0895-6111(00)00059-8
17. Hodge MR, Horton W, Brown T, Herrick R, Olsen T, Hileman ME, McKay M, Archie KA, Cler E, Harms MP, Burgess GC, Glasser MF, Elam JS, Curtiss SW, Barch DM, Oostenveld R, Larson-Prior LJ, Ugurbil K, Van Essen DC, Marcus DS (2016) ConnectomeDB-Sharing human brain connectivity data. Neuroimage 124:1102–1107. https://doi.org/10.1016/j.neuroimage.2015.04.046
18. Fan Q, Witzel T, Nummenmaa A, Van Dijk KRA, Van Horn JD, Drews MK, Somerville LH, Sheridan MA, Santillana RM, Snyder J, Hedden T, Shaw EE, Hollinshead MO, Renvall V, Zanzonico R, Keil B, Cauley S, Polimeni JR, Tisdall D, Buckner RL, Wedeen VJ, Wald LL, Toga AW, Rosen BR (2016) MGH-USC human connectome project datasets with ultra-high b-value diffusion MRI. Neuroimage 124:1108–1114. https://doi.org/10.1016/j.neuroimage.2015.08.075

19. Hore A, Ziou D (2010) Image quality metrics: PSNR vs. SSIM. In: 2010 20th international conference on pattern recognition, pp 2366–2369. IEEE. https://doi.org/10.1109/ICPR.2010.579
20. Dong C, Loy C, He K, pattern XT-I (2015) transactions on undefined: Image super-resolution using deep convolutional networks. https://IEEExplore.ieee.org
21. Bahrami K, Shi F, Rekik I, Shen D (2016) Convolutional neural network for reconstruction of 7T like images from 3T MRI using appearance and anatomical features. Presented at the (2016)
22. Balakrishnan G. Zhao A, Sabuncu MR, Dalca AV, Guttag J (2018) An unsupervised learning model for deformable medical image registration. In: 2018 IEEE/CVF Conference on Computer Vision and Pattern Recognition, pp 9252–9260. IEEE. https://doi.org/10.1109/CVPR.2018.00964

U-control Chart-Based Differential Evolution Clustering for Determining the Number of Clusters in k-Means

Carlos Rondón, Ivon Romero-Pérez, Jesús García Guliany, and Ernesto Steffens Sanabria

Abstract Data are critical sources in several organizations and therefore the efficiency of access to it, sharing, extracting information and making use of that information has become an urgent necessity. Currently, data mining is one of the most recognized fields of research and application for carrying out such tasks. In general terms, data mining is the process of extracting useful information, previously unknown patterns and trends, from large databases. The data mining has received a great impulse in the last times motivated by different causes: (a) the development of efficient and robust algorithms for the processing of large volumes of data, (b) a cheaper computational power that allows the use of computationally intensive methods, and (c) the commercial and scientific advantages that have offered this type of techniques in the most diverse areas (Silva et al in Procedia Comput Sci 151:1219–1224 [1], Dianne and Deborah in Interactive and dynamic graphics for data analysis: with R and Gobi [2]). This paper presents an evaluation from different perspectives of a number of relevant nonsupervised quality measures.

Keywords Clustering · K-means · Particle swarm optimization · Nonsupervised measures

C. Rondón (✉)
Universidad de la Costa (CUC), Barranquilla, Colombia
e-mail: crondon1@cuc.edu.co

I. Romero-Pérez · J. G. Guliany
Universidad Simón Bolívar, Barranquilla, Colombia
e-mail: ivromero@unisimonbolivar.edu.co

J. G. Guliany
e-mail: jesus.garcia@unisimonbolivar.edu.co

E. S. Sanabria
Corporación Universitaria Latinoamericana, Barranquilla, Colombia
e-mail: ssanabria@ul.edu.co

R. Patgiri et al. (eds.), *Proceedings of International Conference on Big Data, Machine Learning and Applications*, Lecture Notes in Networks and Systems 180,
https://doi.org/10.1007/978-981-33-4788-5_7

1 Introduction

In a generic sense, a cluster can be understood as a group or conglomerate of companies and institutions located in the same geographical area, with the purpose of offering similar products and/or services within an industrial sector. This group of companies is characterized by the cooperation between its members, with the aim of increasing international competitiveness and thus contribute to the growth and strengthening of the cluster [1–3].

Within these conglomerates, there are companies that are in direct competition, offering similar products or services. In addition, there may be others offering complementary products and/or services. Also, there may be companies that provide the main inputs for the products and/or services of the cluster [4]. Some terms used in the literature to refer to clusters are [5–7]: industrial clusters, industrial districts, techno-cities and business conglomerations [2]. Although variations are found in the various definitions of a cluster [3], most authors agree on the main characteristics. This characteristic refers to the fact that, due to cooperation and local interaction, the members of the cluster share knowledge that allows them to increase innovation and the development of products and/or services [7]. In this way, the cluster is constantly acquiring and/or developing new technologies to remain at the regional, national or global forefront.

The Particle Swarm Optimization (PSO) algorithm is a stochastic optimization technique that can be used to find an optimal or near optimal solution. It is applied in data and text clustering with very good results [8–10]. PSO can be used to generate good initial centroids for the K-means. In this study, a hybrid PSO + K-means algorithm is used, which combines global PSO search and local K-means search, obtaining improvements in the analyzed quality measures.

2 K-means

The K-means algorithm was proposed by MacQueen in 1968 [11] as a simple, direct algorithm based on the analysis of variances.

The K-means algorithm can be summarized as follows [12]:

1. Selects a random set of initial centroids.
2. Assigns each element of the data set to the nearest centroid.
3. Recalculates the centroids using the following formula

$$c_j = \frac{1}{|C_j|} \sum_{\forall x \in C_j} z \tag{1}$$

   where z represents an element of the data set, which belongs to the cluster C_j; c_j is a centroid and $|C_j|$ corresponds to the number of elements in the cluster C_j.
4. Repeat steps 2 and 3 until the stop condition is reached.

However, if good initial centroids are obtained with an alternative technique, K-means would refine those cluster centroids obtaining better results.

3 Experiments and Results

3.1 Data Set

The algorithms were tested with three types of artificially generated data sets and with three data sets provided by the Donald Bren School of Information and Computer Sciences (http://mlean.ics.uci.edu/MLSummary.html). The artificial data have the following characteristics:

A normal distribution was used and random numbers were generated from the Box–Müller method. Instances with elements of 3, 4, 20, 50 and 100 dimensions were generated in order to be able to translate this data to real problems with large dimensions representing different attributes of an element. In addition to working with data set sizes of 700, 900 and 1200 elements, each data set was formed by four clusters where the elements were distributed in such a way as to obtain three different types of clusters [13]:

- Clusters of equal numbers of elements, either globular or separate (Fig. 1a).
- Clusters of equal numbers of elements, elliptical in shape and intermingled (Fig. 1b).
- Clusters of different number of elements, well globular and separated (Fig. 1c).

Figure 1 shows three-dimensional two artificial data sets, with 1200 elements each (Instances named I26Ba2-1200, I41Bb2-1200, I56Bc2-1200; the name of each instance is formed as follows, e.g., I24Ba50-900 representing the following: I24 instance identifier Ba refers to the distribution of the elements in the type 1(a) data set, 50 is the dimension and 900 is the number of points in the data set). With the characteristics described above, 45 instances were generated, which for the generation of the graphs shown in the following section were grouped as follows [14]:

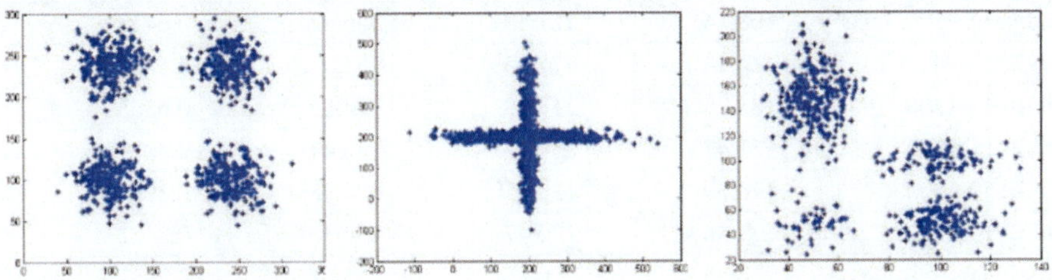

(a) (b) (c)

Fig. 1 Set of three types of artificial data

- Group (a) is formed by 15 instances with the data distributed in the form of Fig. 1a where five instances have dimensions 2, 3, 10, 50 and 100 with 700 elements of the data set each, other five instances have dimensions 4, 20, 50 and 100 with 900 elements of the data set each and the last five instances have dimensions 4, 20, 50 and 100 with 1200 elements of the data set each.
- Group (b) The 15 instances are grouped similarly to group (a) but in this group, the data are distributed as shown in Fig. 1b.
- Group (c) the 15 instances are grouped similarly to group (a) but in this group, the data are distributed as shown in Fig. 1c.

3.2 *Configuration of Algorithms and Results*

For all the algorithms, 30 independent runs were performed. In PSO and PSO + K-means, a cumulus of 15 particles was used for small problem spaces (<10) and for larger dimensions (≥10) the cumulus used was 50 particles. The values of the counters c1 and c2 were set to 1.51 and w = 0.78.

The results of the PSO, K-means and PSO + K-means algorithms are compared below. From 45 artificial instances, three instances were selected from each group and the three instances belonging to the actual dataset.

In Tables 1, 2 and 3, the results obtained in each algorithm of the averages of 30 independent runs for the analyzed instances are shown, where the column: "Instance" corresponds to the chosen instance, "C-Error" shows the quantification of the error, "Intra-C" corresponds to the Intra-Cluster distance, "Inter-C" is the Inter-Cluster distance and the last column, (Intra-Cluster/Inter-Cluster) to the ratio between the two previous measures. Therefore, a good clustering should be one that contains

Table 1 Results obtained with PSO

Instance	C-Error	Intra-C	Inter-C	$\frac{Intra\text{-}c}{Inter\text{-}c}$
I22Ba4-900	37.33	47.14	170.24	0.2475
I23Ba15-900	82.14	88.25	274.36	0.3698
I24Ba50-900	162.85	196.69	625.14	0.3325
I38Bb15-900	168.14	221.47	336.47	0.6458
I39Bb50-900	424.74	474.25	751.5	0.6852
I43Bb15-1200	185.25	214.69	314.65	0.6365
I58Bc15-1200	51.47	56.47	140.47	0.3879
I59Bc50-1200	120.69	120.35	324.98	0.3968
I60Bc100-1200	177.36	170.47	436.36	0.3472
Iris-Plants	0.75	0.85	3.01	0.2958
Glass	1.95	1.67	3.25	0.5347
TAE	12.35	15.32	24.39	0.6587

Table 2 Results obtained with K-means

Instance	C-Error	Intra-C	Inter-C	$\frac{Intra\text{-}c}{Inter\text{-}c}$
I22Ba4-900	32.69	46.35	160.14	0.27581
I23Ba15-900	6825	94.45	247.35	0.32471
I24Ba50-900	160.24	263.24	625.24	0.34587
I38Bb15-900	154.58	242.69	336.98	0.6587
I39Bb50-900	370.14	558.75	774.47	0.73698
I43Bb15-1200	160.69	224.54	396.24	0.62758
I58Bc15-1200	39.47	55.69	114.35	0.37852
I59Bc50-1200	96.36	165.57	335.75	0.42547
I60Bc100-1200	140.96	125.62	469.01	0.48522
Iris-Plants	0.85	0.80	3.46	0.28785
Glass	1.47	2.12	4.47	0.52691
TAE	11.25	13.98	20.08	0.67521

Table 3 Results obtained with PSO + *K-means*

Instance	C-Error	Intra-C	Inter-C	$\frac{Intra\text{-}c}{Inter\text{-}c}$
I22Ba4-900	34.68	48.55	166.96	0.2254
I23Ba15-900	63.44	94.65	297.40	0.2759
I24Ba50-900	142.20	263.84	660.60	0.3142
I38Bb15-900	171.28	242.69	342.85	0.6365
I39Bb50-900	371.90	558.75	768.60	0.6587
I43Bb15-1200	156.63	224.34	435.50	0.4247
I58Bc15-1200	37.47	55.60	150.30	0.3652
I59Bc50-1200	92.21	165.50	342.85	0.3965
I60Bc100-1200	140.32	125.61	470.60	0.3754
Iris-Plants	0.98	0.87	3.32	0.2965
Glass	1.36	2.13	3.47	0.5547
TAE	10.4	13.91	22.68	0.6752

each of the individual homogeneous clusters and among all the clusters as heterogeneous as possible. Where a small value of (Intra-Cluster/Inter-Cluster) represents homogeneity within clusters and heterogeneity between clusters.

It can be seen that in all the cases selected from the artificial data set and for the Iris-Plants instance the PSO + K-means algorithm obtains better results for each of the quality measures analyzed. In these instances, a lower error quantification is obtained, a lower Inter-Cluster distance (more separation) [14] and a lower Intra-Cluster distance (more grouping in the elements), which means that this hybrid algorithm obtains more cohesive and separated clusters between them.

Additionally, Fig. 2 shows the respective values of (Intra-Cluster/Inter-Cluster) of the three algorithms for the group of 15 instances called group (a). It can be seen that with PSO + K-means better results are obtained in 14 of the 15 instances. Figure 3 shows the values of (Intra-Cluster/Inter-Cluster) obtained by the three algorithms for the instances of the group (b). It can be seen that with PSO + K-means better results

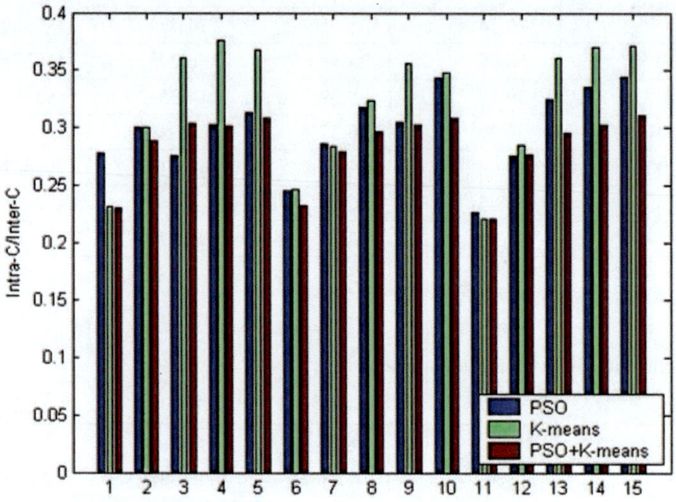

Fig. 2 Values (Intra-Cluster/Inter-Cluster) for the group a

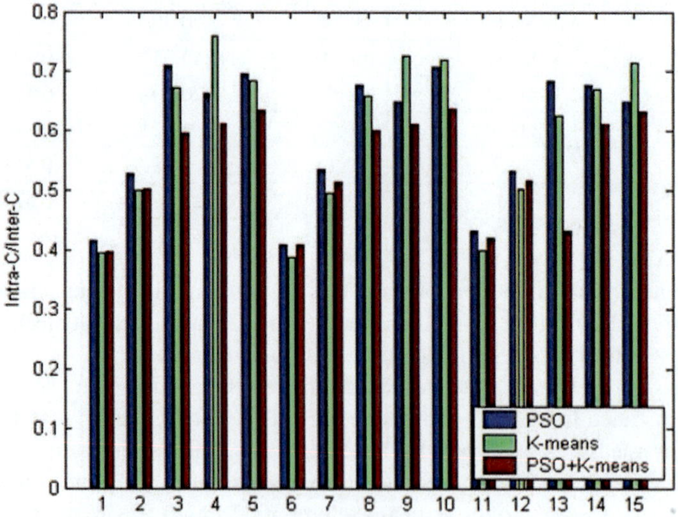

Fig. 3 Values (Intra-Cluster/Inter-Cluster) for the group b

are obtained in 9 of the 15 instances. Figure 4 shows the values (Intra-Cluster/Inter-Cluster) of the three algorithms for the instances in the group (c). In this case, it can be seen that with this distribution of data with the PSO + K-means algorithm, better results are obtained in each and every one of the 15 instances. Finally, for the real data set, Fig. 5 shows the best results in two of the three instances.

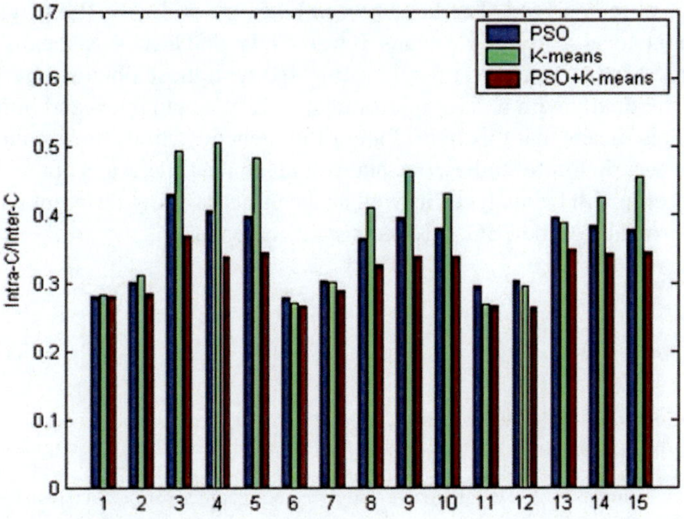

Fig. 4 Values (Intra-Cluster/Inter-Cluster) for the group c

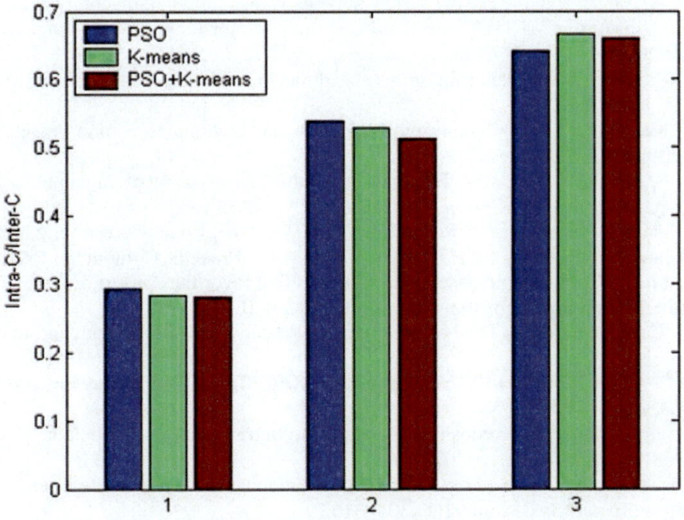

Fig. 5 Values (Intra-Cluster/Inter-Cluster) for the actual data set, where 1 = Iris, 2 = Glass y 3 = TAE

4 Conclusions

PSO is an optimization technique that performs a global search for solutions and can be applied to clustering problems. K-means is a simple, direct and widely used algorithm in clustering but it can get stuck in local optimizations. In this paper, quality measures of clusters obtained by three algorithms: PSO, K-means and PSO + K-means, were analyzed. The last one combines the ability of PSO's globalized search with the local search of K-means. It was shown that PSO + K-means was able to find, in most cases, two better results compared with those obtained by K-means and PSO individually with regard to error quantification, and inter- and intra-cluster distances. This means that this hybrid algorithm generally finds more cohesive and separate clusters. In future studies, the quality and number of clusters obtained by this hybrid algorithm will be analyzed as well as the effect of using different parameters on the performance of the PSO + K-means convergence.

References

1. Silva J, Cubillos J, Villa JV, Romero L, Solano D (2019) Preservation of confidential information privacy and association rule hiding for data mining: a bibliometric review. Procedia Comput Sci 151:1219–1224. R
2. Dianne D, Deborah F (2007) Interactive and dynamic graphics for data analysis: with R and Gobi. ISBN: 978-0-387-71762-3
3. Fraley C, Raftery A (2007) Model-based methods of classification: using the mclust software in chemometrics. University of Washington Seattle, United States.https://doi.org/10.18637/JSS.V018.I06
4. Viloria A, Viviana Robayo P (2016) Virtual network level of application composed IP networks connected with systems—(NETS Peer-to-Peer). Indian J Sci Technol 9(46). https://doi.org/10.17485/ijst/2016/v9i46/107376.
5. Witten I, Frank E (2000) Data mining: practical machine learning tools with Java implementations. M. Kaufmann
6. Jlawe H, Kamber M (2002) Data mining: concepts and techniques. Simón Fraser University. Morgan Kaufmann Publishers
7. Moreno A, Redondo T (2016) Text analytics: the convergence of big data and artificial intelligence. Int J Interact Multimed Artif Intell
8. Silva J, Solano D, Fernandez C, Romero L, Villa JV (2019) Privacy preserving, protection of personal data, and big data: a review of the Colombia case. Procedia Comput Sci 151:1213–1218
9. Heaton J (2016) Comparing dataset characteristics that favor the Apriori, Eclat or FP-Growth frequent itemset mining algorithms. In: SoutheastCon. IEEE, pp 1–7
10. Bennet J, Lanning S (2007) The Netflix prize. In: Proceedings of KDD cup and workshop, pp 3–6
11. Linyuan L, Matúš M, Chi Ho Y, Yi-Cheng Z, Zi-Ke Z, Tao Z (2012) Recommender systems. Elsevier B.V, pp 1–49
12. López Puga JG (2007) Las redes bayesianas como herramientas de modelado en psicología. Anales de Psicología. Redalyc, 307–316
13. Taneja A, Arora A (2018) Cross domain recommendation using multidimensional tensor factorization. Expert Syst Appl 92(1):304–316
14. Huang TW (2006) Application of clustering analysis for reducing SMT setup time—a case study on Avantech Company. Master's thesis, Department of National Taipei University of Technology

15. Witten IH, Frank E (2005) Data mining: practical machine learning tools and techniques, 2nd edn. Morgan Kaufmann series in data Management Systems. ISBN: 0-12-088407-0
16. Chui C-Y, Chen Y-F, Kou I-T, Ku HC (2008) An intelligent market segmentation system using k-means and particle swarm optimization. Expert Syst Appl
17. Yue W (2016) Research on the clustering analysis algorithm for data mining. Revista Ibérica de Sistemas y Tecnologías de la Información (RISTI), E6, pp 209–221
18. Zhu Z (2015) A clustering method for high-dimensional data analysis in stock market. Revista Ibérica de Sistemas y Tecnologías de la Información (RISTI), 17A, pp 209–221
19. Keim D, Kohlhammer J, Geoffrey E, Mansmann F (2010) Mastering the information age solving problems with visual analytics. Edited by the authors Published by the Eurographics Association Postfach 8043, Printed in Germany, Druckhaus Thomas Müntzer GmbH, Bad Langensalza. Theoretical Issues in Ergonomics Science, vol 8, no 1. ISBN 978-3-905673-77-7
20. Lee H, Kihm J, Choo J, Stasko J, Park, H (2012) iVisClustering: an interactive visual document clustering via topic modeling. In: Computer graphics forum, vol 31, no 3pt3. Blackwell Publishing Ltd., pp 1155–1164
21. Silva J, Varela N, Lezama OBP, Hernández-P H, Ventura JM, de la Hoz B, Coronel LP (2019) Multi-dimension tensor factorization collaborative filtering recommendation for academic profiles. In: International symposium on neural networks. Springer, Cham, pp 200–209, July 2019
22. Kim J, Ale J (2004) Descubrimiento incremental de las reglas de asociación temporales. X Congreso Argentino de Ciencias de la Computación
23. Castillo W, Meneses C (2012) A comparative review of schemes of multidimensional visualization for data mining techniques. In: III Congreso Internacional de Computación e Informática del Norte de Chile (INFONOR-CHILE), Arica, Chile

Novel Initialization Strategy for K-modes Clustering Algorithm

Aizan Zafar and K. Swarupa Rani

Abstract K-modes is a method that is computationally efficient to cluster categorical data. To perform a better clustering, an appropriate selection of initial modes is required by K-modes clustering algorithm. Hence, selecting an initial mode for K-mode clustering algorithm plays a crucial role in the performance of clustering. Currently, for numerical data, different initialization methods exist, but we cannot use those initialization methods for categorical data due to lack of geometry. In this paper, we proposed and adopted density parameter-based method for selecting an initial cluster mode integrated with the traditional K-modes clustering algorithm, leading to more accurate cluster results. Our proposed method selects an initial cluster mode so that it will be closer to the final cluster mode. The experiment has been carried out on different benchmark datasets and compared with the traditional methods.

Keywords K-mode clustering · Average similarity · Density parameter

1 Introduction

We are living in a world full of data. Storage of a huge amount of data and to represent the data is a challenging task for analysis and management. One can either classify or cluster these types of data to handle them effectively. Clustering is an unsupervised learning method, this method partitions into groups. The philosophy of clustering is to form groups based on similarity and dissimilarity measure for data objects [1]. The reason for clustering is significant in numerous fields, for example, Image Processing, Pattern Recognition, Data mining, Information Science, etc.

A. Zafar · K. Swarupa Rani (✉)
School of Computer and Information Sciences, University of Hyderabad, Hyderabad 500046, India
e-mail: kswarupaprasad@gmail.com

A. Zafar
e-mail: aizanzafar@gmail.com

R. Patgiri et al. (eds.), *Proceedings of International Conference on Big Data, Machine Learning and Applications*, Lecture Notes in Networks and Systems 180,
https://doi.org/10.1007/978-981-33-4788-5_8

Major clustering methods have existed in the literature. Categorization of these methods are difficult due to the overlap of several categories. But, in general, these are classified into two major categories. (i) Partitional clustering approach, constructs "m" partitions represent a cluster. Few popular heuristic methods in partitioning clustering are K-means and K-medoids algorithm. (ii) Hierarchical clustering approach creates a hierarchical decomposition of the data. It further classifies it as an agglomerative and divisive approach.

Many researchers worked with basic K-means [2] clustering algorithm which is suitable for numerical data. Ralambondrainy et al. [3] experimented with the same K-means algorithm by transforming the numerical data to categorical data form and obtained the incorrect results and the same has been reported by their observations and analysis [3].

However, the experiments continued and carried out by many researchers to handle categorical data. K-modes algorithm which has been discovered by Huang [4] used for clustering the categorical data by extending and modification of traditional K-means clustering algorithm. This algorithm, which can cluster categorical data introduces a new dissimilarity measure by replacing means of the cluster with the value of the most frequent attribute value. K-modes clustering algorithm has some limitations. Determine the number of clusters, Identifying outliers, handling empty clusters are some of the issues. The main challenge is to select the initial modes from the dataset. To tackle this problem, Zhu et al. [5], used the density parameter to select initial centroids for a numerical dataset.

This motivates us to use the same density parameter of each data point to function with categorical data so that the initial modes are no longer selected randomly. Formation of clusters (groups) based on the similarity measures is an important and challenging phases in various applications. The partitional clustering approach works by choosing an initial data object from random selection. This approach is good when the number of clusters is less and based on the selection of the good random initialization objects.

One of the main challenges is to choose an initial center which is having direct impact on the formation of clusters. Fuyuan et al. [6] and Wu et al. [7] come up with a new initialization method for choosing a cluster center to categorical data. But, these researchers defined the density of an object in two different ways. Bai et al. [8] proposed different techniques for initializing categorical data. This approach is based on the density and the distance function to compute the initial center. Bai et al. [8] define the density of an object by calculating the average distance of all the other point from a point. Khan et al. [9] observed and analyzed the accuracy, recall, and precision values are incorrect for some datasets presented in the works of Bai et al. [8]. To overcome the limitations of Bai [8], Khan [9] proposed three different attribute selection methods for categorical data. Sajida et al. [10] uses separation parameters to a method to select the initial data object from a dense region. Zhu et al. [5], introduces three components to numerical data such as (i) a novel approach for initializing cluster center, (ii) a new approach variance clustering validity index to evaluate the clusters, and (iii) an improved K-means algorithm to form a cluster.

Table 1 Survey of various clustering technique from [1998–2018]

S.No	Technique	Clustering type	Advantage
1	K-modes, Huang [4]	Categorical data	To handle categorical data used a simple matching dissimilarity measure to deal with it
2	New initialization method for categorical data, Cao et al. [6]	Categorical data	Better than random initialization and defined density of object based on frequency of attributes
3	Cluster center initialization method, Bai et al. [8]	Categorical data	The distance between points and an object's density is used
4	Initial cluster center using multiple attributes clustering, Khan et al. [9]	Categorical data	Selecting the initial cluster core by multiple data clustering based on the present value of the attributes
5	Seed selection method, Sajda et al. [10]	Categorical data	Uses the density estimation of data points to compute distance of each data point
6	Improved K-means and Variance based clustering validity index, Zhu et al. [5]	Numerical data	Proposed a density parameter for initializing K-means centers and VCVI to find optimal number of cluster

We proposed in this paper, by integrating the idea of Zhu [5] initial cluster center with K-modes algorithm [4] to form a cluster to categorical data. We used the density parameter based initial centers [5] to K-mode algorithm [4] by using matching dissimilarity measures to form clusters by updating the modes of a cluster in each iteration. According to Bezdek [11], we used Empirical rule, i.e., $K_{max} \leq n$ where "n" represents the number of features in dataset D. The experiment has been carried out for the proposed approach for different benchmark datasets (Table 2).

This paper's description is as follows. In Sect. 2, we briefly introduce K-mode clustering algorithms. In Sect. 3, a proposed approach for categorical data has been introduced. In Sect. 4, Experimental results have been evaluated. Finally, Sect. 5 we briefly concludes this paper.

2 K-mode Algorithm for Categorical Data Clustering

Huang [4] discovered the K-mode algorithm by modifying the existing K-mean algorithm [2] by using (i) dissimilarity measure to match for the categorical object

(ii) by using modes instead of cluster means. The mode will be selected based on the frequency of the value in an attribute.

The obstruction in the k-means clustering technique is removed through a few adjustments. These modifications are

- For categorical data using a simple corresponding measure of dissimilarity.
- Use modes as a cluster center instead of means.
- Find the mode using a frequency-based method.

2.1 *Dissimilarity Measure*

Suppose we have two categorical objects (say P and Q), which are described by “n” categorical attributes. We can define d(P, Q) as a dissimilarity measure between P and Q as total number of discrepancies for two objects matching attributes.

Mathematically we can say that

$$d(P, Q) = \sum_{j=1}^{n} \delta(p_j, q_j) \tag{1}$$

where,

$\delta(p_j, q_j) = 1, if p_j \neq q_j$ or, 0 otherwise

2.2 *Mode of a Set*

Suppose we have a categorical object (say Y) described by “m” categorical attributes, $C_1, C_2, \ldots, C_m$. Mode of $\mathbf{Y} = [Y_1, Y_2, \ldots, Y_n]$ is a vector R = $[r_1, r_2, \ldots, r_m]$ that minimizes

$$D(Y, R) = \sum_{i=1}^{n} d(Y_i, R) \tag{2}$$

Here, R is not necessarily an element of **Y**.

2.3 *Cost Function*

Let “M” be a set of “m” categorical data objects where each data object is described by the help of “n” categorical attributes,$N_1, N_2, \ldots, N_n$. When the dissimilarity measure for categorical data defined in Eq. (1) is used for the categorical data object in set M, the cost function becomes

$$C(Q) = \sum_{i=1}^{m} d(M_i, Q_i) \tag{3}$$

where M_i is the ith data object and Q_i as closest cluster center for that data element.

2.4 *The K-mode Algorithm*

The traditional steps used in K-mode algorithm is given below

1. "k" initial mode is selected for "k" number of clusters.
2. Assign the data objects to the cluster according to Eq. 3. and update cluster mode.
3. At the end of cluster formation, verify the modes using dissimilarity measure. If any object is close to the other mode of a cluster then reallocate the object and update the mode
4. Continue step 3, until no objects have reallocated to a different cluster

3 Our Proposed Approach for Effective Clustering

We discussed different approaches for categorical data clustering in Sect. 1. In traditional K-modes algorithm [4], the modes for clusters center are randomly selected. Therefore, it may lead to the problem of slow convergence of clustering due to random selection. To boost K-modes algorithm [4] efficiency and to overcome the slow convergence of clustering, we proposed an approach by integrating K-modes algorithm with modified density parameter (Average Dissimilarity AD measure) introduced by Zhu [5] and the details are explained below.

3.1 *Average Dissimilarity*

In a dataset, D = $[x_1, x_2, \ldots, x_n]$ the Average Dissimilarity (AD) is computed between all the data points pair (x_i, x_j) in dataset D, is defined as

$$AD = \frac{2}{n * (n-1)} \sum_{i=1}^{n-1} \sum_{j=i+1}^{n} d(x_i, x_j) \tag{4}$$

where $d(x_i, x_j)$ is the dissimilarity measure used in Eq. 1.

3.2 Neighborhood Value

A dataset having “n” object, D = $[x_1, x_2, \ldots, x_n]$ is given. When we select each sample point in the fields $x_i (i = 1, 2, \ldots, n)$ as a center then its neighborhood value should be defined in such a way that average dissimilarity in a dataset D should be within threshold value. The neighborhood value of each data point in a dataset D is defined as

$$AvgDis = (AvgDis - \alpha) \tag{5}$$

where α is known as an influence factor of neighborhood value.

3.3 Density Parameter

A dataset having “n” object, D = $[x_1, x_2, \ldots, x_n]$ is given. The density parameter is defined for each of data objects in the fields when $x_i (i = 1, 2, \ldots, n)$ is taken as a center and AD used as a maximum value requirement for that sample point to come under its neighborhood. Mathematically, Eq. (6) determines the density parameter and the corresponding jump function is also defined by Eq. (7).

$$\rho(x_i, AD) = \sum_{i=1, j \neq i}^{n} sgn(d(x_i, x_j)) \tag{6}$$

where jump function sgn is defined as

$$sgn(x) = 1, if(d(x_i, x_j) \leq AD), otherwise0 \tag{7}$$

From the above definitions, we have proposed an Improved K-modes algorithm which is given below in Algorithm 1.

Algorithm 1 Improved K-mode Algorithm

Input: Dataset D = {x_1, x_2, ..., x_n} & Number of cluster(k)
Output:k cluster are divided into C = {C_1, C_2, ..., C_k}
1. Calculate the Average dissimilirarity between all the data point (x_i, x_j) in D using Eq. 4
2. Let $S_i = \phi(1 \leq i \leq n)$
for *i=1 to n* **do**
  3. S = {$\rho(x_i, AD)$}, (Calculate Density parameter $\rho(x_i, AD)$ of each sample point x_i using the Eqs. (6) and (7))
  4. Store it into set S
end
for *j = 1 to k* **do**
  5. Select jth largest density parameter value,(x_j=max{S}), represent initial cluster mode v_j
  6. Allocate the sample points from dataset D based on cluster mode v_j whose dissimilarity measure is less than or equal to AD
  7. Remove allocated data objects from set S
end
8. The initial cluster mode sets is:{v_1,v_2,...,v_k}
9. Repeat
10. Let $C_i = \phi(1 \leq i \leq K)$
for *j = 1 to n* **do**
  11. Calculate the dissimilarity score between sample point x_j to all the cluster center v_i($1 \leq i \leq K$)
  12. Allocate sample point x_j to cluster whose score is minimum
  13. Use "Moving Mode Frequency Based Method" to update the cluster mode
end
14. until modes are converge

The following are the description of the Improved K-modes algorithm:

In step 1, computing the average dissimilarity of the sample points according to the Eq. (4). In steps 3–5, calculating the density parameter according to the Eqs. (6) and (7). In steps 6–9, choosing the sample point which has the highest density value as one of the initial cluster modes. With the initial cluster mode and also considering density parameter, some of the data points form a cluster and are removed from set S. From the remaining data points in set S, the next highest value is selected as a second initial cluster modes and again will repeated the same process until we get predefined number of cluster center. In step 10, we have the initial cluster modes value for the given dataset. Steps 11–16, is a traditional K-mode algorithm applied for the above calculated initial cluster center.

In this way, the improved K-modes clustering method selects initial cluster center points. By this proposed approach, we observed stable clustering results.

4 Experimentation and Analysis

We present the experimental study to check the performance of the proposed method on a few standard datasets from the UCI Machine Learning Repository. The specifications of the system, in which this algorithm are implemented, is an Intel R CoreTM i5-7500U CPU @ 3.40GHz. The memory size (RAM) is 4GB and the disk space is 500 GB. We have implemented our algorithms in Python using Jupyter Notebook version 5.6.0.

4.1 Evaluation Method

Using an extrinsic evaluation of cluster, we want to compare our clustering results with the ground truth values, i.e., actual class of cluster.

Our proposed, an Improved K-modes clustering algorithm has been evaluated with the indices introduced by Wu [7], i.e., Accuracy (AC) and Precision(PR) can be defined as

$$AC = \frac{\sum_{i=1}^{k} e_i}{N} \tag{8}$$

where e_i denote the data object which is accurately nominated for class C_i.

$$PR = \frac{\sum_{i=1}^{k} \frac{e_i}{e_i + b_i}}{K} \tag{9}$$

where e_i denote the data object which is accurately nominated for class C_i and b_i denotes data object which is wrongly nominated for class C_i

4.2 Dataset

We have experimented on 4 different standard categorical datasets from the UCI Machine Learning Repository. In Table 2, we have shown the description of dataset. Table 3, shows the comparison result of our proposed algorithm with 5 different exiting algorithms.

The description of Tables 4, 5, 6 and 7 are as follows:

The class represents some objects in a predicted cluster on the benchmark dataset. The accuracy is tabulated in Table 4 based on the proposed methodology of forming of the cluster and the actual cluster values in the second column.

Table 2 The description of the dataset

Datasets	Sample object	Desired cluster	Number of dimensions	'K' range
Dermatology	366	6	34	[2, 19]
Soyabean	47	4	35	[2, 7]
Zoo	101	7	18	[2, 10]
Lung-Cancer	32	3	57	[2, 6]

Table 3 Comparative study

Dataset	Measure	Random K-mode	Wu et al. [7]	Cao et al. [6]	Khan et al. [9]	Sajida et al. [10]	Proposed
Soybean	AC	0.8644	1	1	0.9574	1	**1**
	PR	0.899	1	1	0.9583	1	**1**
Dermatology	AC	0.6870	–	0.7486	0.7372	0.6885	**0.7869**
	PR	0.7633	–	**0.8801**	0.7909	0.6293	0.8050
Zoo	AC	0.8324	0.8812	0.8812	0.8911	0.7327	**0.9109**
	PR	0.8186	0.8702	0.8702	0.7224	0.6077	**0.8768**
Lung-Cancer	AC	0.5363	0.500	0.500	0.500	0.5625	**0.6875**
	PR	0.6033	0.5584	0.5584	0.6444	0.5766	**0.7222**

Table 4 Cluster evaluation for Dermatology dataset

Cluster ID	Actual cluster	Predicted clusters with different classes					
		Class1	Class2	Class3	Class4	Class5	Class6
C1	49	44	0	5	0	0	0
C2	52	5	45	0	0	0	0
C3	61	24	0	37	0	0	0
C4	112	4	2	0	72	0	27
C5	72	4	0	4	0	64	0
C6	20	0	2	1	0	0	17

4.2.1 Dermatology Dataset

This dataset contains 366 sample points, each sample point is represented by 33 dimensions or attributes. The whole dataset is partitioned into six clusters. The clustering result of our proposed Improved K-mode algorithm is summarized in Table 4.

Table 5 Cluster evaluation for Zoo dataset

Cluster ID	Actual cluster	Predicted clusters with different classes						
		Class1	Class2	Class3	Class4	Class5	Class6	Class7
C1	41	34	3	0	0	0	4	0
C2	8	0	8	0	0	0	0	0
C3	20	0	0	20	0	0	0	0
C4	14	1	0	0	13	0	0	0
C5	4	0	1	0	0	3	0	0
C6	4	0	0	0	0	0	4	0
C7	10	0	0	0	0	0	0	0

Table 6 Cluster evaluation for Soybean dataset

Cluster ID	Actual cluster	Predicted clusters with different classes			
		Class1	Class2	Class3	Class4
C1	10	10	0	0	0
C2	10	0	10	0	0
C3	10	0	0	10	0
C4	17	0	0	0	17

4.2.2 Zoo Dataset

This dataset consists of 101 sample points, each sample point is represented by 18 attributes. The whole dataset is partitioned into seven different classes. The clustering result of our proposed Improved K-mode algorithm is summarized in Table 5.

4.2.3 Soybean Dataset

This dataset contains 47 sample points, where each sample point is represented by 35 dimensions or attributes and they are categorized as one of the 4 diseases. The clustering result of our proposed Improved K-mode algorithm is summarized in Table 6.

4.2.4 Lung-Cancer Dataset

This dataset consists of 32 records, each represented by 56 categorical attributes. The dataset has been divided into three different classes. The result of the proposed Improved K-mode algorithm is shown in Table 7.

Table 7 Cluster evaluation for Lung-Cancer dataset

Cluster ID	Actual cluster	Predicted cluster		
		Class1	Class2	Class3
C1	10	10	3	0
C2	13	2	4	4
C3	9	0	1	8

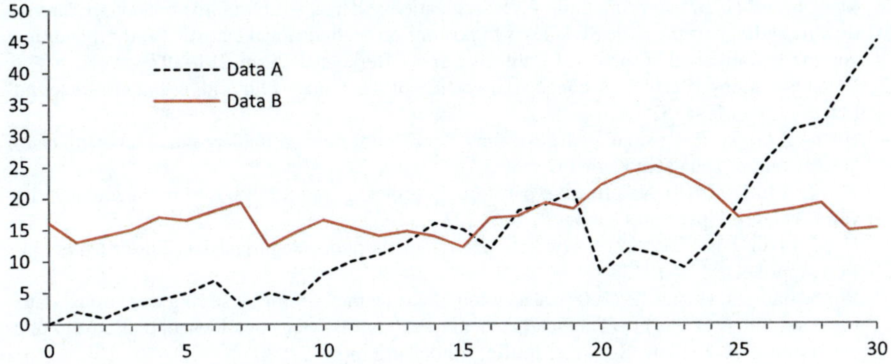

Fig. 1 Computational times for different numbers of clusters

4.2.5 Experimentation to Run Time of Proposed Approach

We observed the run time of our proposed method on various datasets, which tends to increase linearly about to the number of clusters, which is shown in Fig. 1.

5 Conclusions and Future Work

Now a day, categorical data is present in real-world databases. This leads to the development of K-mode clustering algorithm. However, the clustering algorithm needs to re-run multiple times with different initialization to find a good solution. Therefore, it works well when we have a small dataset, so that there are good chances of getting at least one random cluster center near the final cluster solution. In this paper, we came up with a new initialization method for the cluster center modes based on the density parameter, an efficient and effective method. Furthermore, it overcomes the shortcomings of the existing initialization methods in terms of accuracy but at the expense of time efficiency for a large dataset. We have demonstrated our methods with four separate categorical data sets from the UCI Machine Learning Repository and experimental results have been presented. One can enhance this algorithm to incorporate this method to handle real-world datasets and heterogeneous datasets.

References

1. Han J, Pei J, Kamber M (2011) Data mining: concepts and techniques. Elsevier
2. MacQueen J (1967) Some methods for classification and analysis of multivariate observations. In: Proceedings of the fifth Berkeley symposium on mathematical statistics and probability, volume 1: statistics. University of California Press, Berkeley, CA, pp 281–297
3. Ralambondrainy H (1995) A conceptual version of the k-means algorithm. Pattern Recognit Lett 16(11):1147–1157
4. Huang Z (1998) Extensions to the k-means algorithm for clustering large data sets with categorical values. Data Min Knowl Discov 2:283–304
5. Zhu E, Ma R (2018) An effective partitional clustering algorithm based on new clustering validity index. Appl Soft Comput 71:608–321
6. Cao F, Liang J, Bai L (2009) A new initialization method for categorical data clustering. Expert Syst Appl 36:10223–10228
7. Wu S, Jiang Q, Huang JZ (2007) A new initialization method for clustering categorical data. In: Proceedings of the 11th Pacific-Asia conference on advances in knowledge discovery and data mining, PAKDD07. Springer, Berlin, Heidelberg, pp 972–980
8. Bai L, Liang J, Dang C, Cao F (2012) A cluster centers initialization method for clustering categorical data. Expert Syst Appl 39(9):8022–8029
9. Khan SS, Ahmad A (2013) Cluster center initialization algorithm for K-modes clustering. Expert Syst Appl 40(18):7444–7456
10. Sajidha SA, Chodnekar SP, Desikan K (2018) Initial seed selection for K-modes clustering—a distance and density based approach. J King Saud Univ Comput Inf Sci
11. Bezdek JC (1998) Pattern recognition in handbook of fuzzy computation. IOP Publishing Ltd.

A Survey on Streaming Data Analytics: Research Issues, Algorithms, Evaluation Metrics, and Platforms

D. Christy Sujatha and J. Gnana Jayanthi

Abstract Data is the most significant element for any organization in today's cut-throat competition, which is growing tremendously large and exchanged over the Internet due to the development of numerous digital devices. Commercial firms are focusing on predicting responses in advance before actually spending valuable resources. Nowadays most of the companies are moving from traditional static data analytics to real-time streaming data analytics to renovate the predictive model to be adaptable with currently available data which changes over time. Although building a predictive model for the real-time streaming data is an immense challenge, it is a current area of Research and Development which gained much attention from the Academia, IT Industries, and Researchers. In this survey paper, we have instigated with discussing some of the open research issues of streaming data and then we deliberate several classification algorithms proposed by various researchers by analyzing it's merits and limitations. Besides, we studied the evaluation metrics used by the researchers for the purpose of measuring the existing algorithms. Finally, we surveyed the platforms used by the researchers to handle streaming data and summarized a comparative study of streaming data platforms.

Keywords Streaming data · Challenges · Classification algorithms · Platform · Metrics

1 Introduction

Big data is a big buzzword these days, for the reason most of the data is born in the form of digital nowadays, growing tremendously large and exchanged over the Internet. According to the most recent statistics, the world population is 7.6

D. Christy Sujatha · J. Gnana Jayanthi (✉)
PG and Research Department of Computer Science, Rajah Serfoji Government College, Thanjavur, Tamil Nadu, India
e-mail: jgnanajayanthi@gmail.com

D. Christy Sujatha
e-mail: christy_se@pmu.edu

R. Patgiri et al. (eds.), *Proceedings of International Conference on Big Data, Machine Learning and Applications*, Lecture Notes in Networks and Systems 180,
https://doi.org/10.1007/978-981-33-4788-5_9

billion as of June 2018 and it is expected to reach 8 billion people in 2023 [1]. These billions of people are generating a massive amount of data through the increased use of mobile devices, social networks, remote sensors, telecalls continuously produce much heterogeneous data, either in the form of structured or unstructured data [2].

Merely collecting and storing raw data is unusable and it seems to be only noise unless the hidden information is obtained from the huge data. Data is the most vital element for any organization in today's cut-throat competition. Commercial firms are focusing on predicting responses in advance, before actually spending valuable resources for building an effective decision-making. It is possible to get a tremendous amount of valuable information, hidden pattern, trends, and associations between data set from big data analytics [3, 4]. Predictive analytic models can be built by using Statistical algorithms and Machine Learning algorithms to identify the possibility of future outcome based on historical data [5].

In conventional static data model, the entire data is already stored over a period of time which will undergo processing later. Unlike traditional static data sets, streaming data flow in and out of a system continuously with varying data distribution [6]. It is an instant analytics of real-time data as soon as the data is generated. Nowadays more companies are moving from traditional batch data analytics to real-time streaming data analytics as we could get the information very quickly. Building a predictive model for the real-time streaming data is undoubtedly an immense challenge [7] and there is a lack of support and resources for streaming data. Hence, it is a current area of research and development which gained much attention from the Academia, IT Industries, and Researchers [8].

This survey paper is an overview of streaming data analytics challenges, existing classification algorithms, evaluation metrics, and suitable platforms to handle streaming data. Section 2 describes the research approach questions related to streaming data and Sect. 3 explains various open research issues. Section 4 deliberates several Classification algorithms proposed by several researchers for streaming data. Section 5 shows the performance metrics used by the researchers to evaluate the predictive model. Section 6 deals with the comparative study of platforms to handle streaming data. Section 7 concludes with the scope and future of streaming data analytics.

2 Research Approach

This survey paper presents a detailed analysis of the following research approach queries to comprehend the existing research proposals on Streaming data.

Research Questions:

RQ1: What are the research issues and challenges of Steaming Data Analytics?
RQ2: How the challenges are overcome by classification algorithms/techniques proposed by various researchers?

RQ3: What are the evaluation metrics for the predictive model used by various researchers?

RQ4: What are the suitable platforms to perform streaming data analytics?

Based on the above research questions, a systematic analysis review is performed by considering various research papers and articles spanning from the year 2000 up to 2018, published by ACM, Elsevier, Springer, IEEE, and other Journals.

3 RQ1: Research Issues of Streaming Data Analytics

Data stream has several challenges to resolve and the researchers in the surveyed articles focused on the following problems.

3.1 Concept Drift

According to data mining, *Concept* refers to the target information or variable that any predictive algorithm is trying to predict. There is always a relationship between input features and output target variable. This relationship is not static and changes over time in streaming data, which yields the challenge of concept *drift*. Most of the researchers focused on the problem of Concept drift [9–22]. Since the analytics model developed at a particular timestamp may not be suitable for the continuously incoming stream of data, most of the authors proposed an adaptable model toward the new incoming data which dynamically changes over time.

3.2 Resource Optimization

In this literature, it is observed that some of the authors have focused on resource optimization which is usually associated with the mathematical and machine learning part of data science. Almost most of the authors tried to minimize the resources like *Memory* [23, 24] *and Time taken* for training, learning, and execution [25–29]. It is not only enough to make predictions or conclusions from the available data set, but making a decision or choosing the best solution is also an important issue. The authors Linande Li et al. and Victor Guilherme et al. [23, 24] have proposed that behaviors of each resource is directly proportional to one another, e.g., reducing memory size automatically reduces the cost and time taken by the model.

3.3 Distributed and Parallel Systems

Distributed and Parallel system computes on multiple concurrently used computing units which may be of different cores of the same processor or different processors, or even single core with emulated concurrent execution. A. Anandha Babu et al. [27] handled streaming data using Apache SPARK but not clearly described the parallel processing work. Sharmishta Desai et al. [28] and Pranmitha Nanda et al. [30] implemented HADOOP for the distributed and parallel approach to reduce the computational and time complexity. But Hadoop Map Reduce is not a suitable platform for streaming data as it slows down the execution.

3.4 Overfitting/Underfitting

A too complex model with high variance leads to overfitting and too simple model/with high bias leads to underfitting problem. *Overfitting* is an error in a model which occurs when a function of the model is too closely fit the set of data points. *Underfitting* occurs when the model does not fit the data well enough. Such a model will lead to poor predictive performance. To get more accuracy in prediction, the data set plays a vital role in a learning model. But this major problem is addressed by very few researchers Hang Yang et al. [31] and Jihed Kiari et al. [32].

3.5 Class Imbalance

Very few researchers, Chen Li et al. [33], Hang Yang et al. [31], and Anand Gupta et al. [34] handled the problem of class imbalance. *Class imbalance* is the problem in machine learning where there is more variation between the total number of one set of classes with another set of classes and hence it is challenging to build a model for class imbalanced data set. They argued that little imbalance is not a problem and the problem arises when the class imbalance is high. This situation arises due to the fact that the minority class may be rare or difficult to obtain. Though highly imbalanced classes have more accuracy, good insights may not be generated and the minority classes will have low recall value.

3.6 Other Research Issues

Apart from the above research issues, other researchers proposed their ideas to solve the problems such as (i) to handle heterogeneous data [35], (ii) to overcome the dependency of the number of examples [36], (iii) to select the best classifier [37, 38],

(iv) to preserve previous model [39], (v) to learn from multiple class [40], (vi) to label extraction from the current model [41], (vii) to find k value from the training data [42], and (viii) Histogram-based classification [20].

Figure 1 shows the proportion of research problems handled by various researchers. From the survey, it is perceived that most of the researchers handled the problem of Drift Detection, Drift Prevention, Drift Adaptation, Memory, and Time Minimization. Some researchers applied Distributed and Parallel Systems to speed up the analytics process. Moreover, it is also observed that Class Imbalance and Data Overfitting problems are addressed by a lesser amount of researchers which may be considered as an open research problem to be focused on in future. Because overfitting and class imbalance are the major issues affecting the accuracy of prediction (Fig. 2).

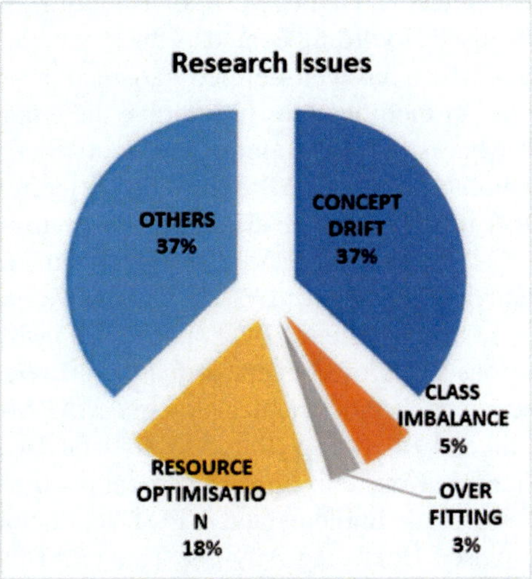

Fig. 1 Research issues of streaming data

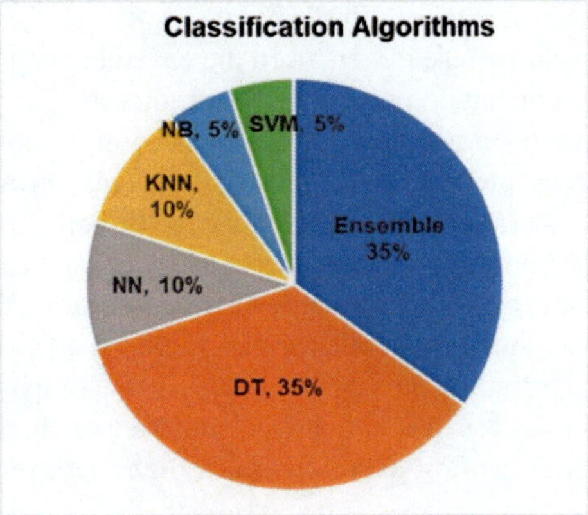

Fig. 2 Classification algorithms for streaming data

4 RQ2: Classification Algorithms for Streaming Data

Many algorithms have been developed for streaming data analytics using Classification, Regression and Clustering. In this survey, we analyze only the classification algorithms which come with user-supervised machine learning. Classification is the process of predicting the class of any unknown instance, based on the constructed model. Streaming data cannot be split into training and testing data as we used for traditional static data analytics since the entire data cannot be stored at the beginning, rather the incoming examples are trained and tested either instance by instance or batch of instances. Table 1 shows the summary of existing classification algorithms proposed by various researchers who used the standard algorithms like Decision tree, KNN, NN, SVM, Naive Bayes, and Ensemble approach.

From the survey, it is observed that most of the algorithms are based on Very Fast Decision Tree (VFDT) or Hoeffding Tree (HT) proposed by Domingos et al. who used Hoeffding Bound as the number of examples to be observed for streaming data, which has well-defined statistical proof [43] and many researchers proposed a series of modifications to improve the predictive performance and to overcome the drawbacks of VFDT algorithm. Geoff Hulten et al. followed VFDT and proposed a new algorithm CVFDT to convert the Hoeffding Tree to be adaptable with the concept Drift in a dynamic distributed data environment [9]. Joao Gama et al. introduced $VFDT_c$ by enhancing VFDT to deal with continuous (or) numerical data [35]. Ruth Ramya et al. enhanced VFDT with Naive Bayes at the leaf node [25], Gang Liu et al. introduced Efficient VFDT (E-CVFDT) with cache mechanism in VFDT to identify various types of concept drift [10]. Mirela Teixeira Cazzolato et al. proposed an incremental statistical decision tree StARMiner Tree (ST), replaced the Information Gain and Hoeffding Bound by statistical values mean and standard deviation [36]. Nicholas Kourtellis applied vertical parallelism in Hoeffding Tree [26]. Sharmishta Desai et al. implemented VFDT on Hadoop [28] Chaitanya Manapragada et al. proposed Hoeffding Anytime Tree (HATT) algorithm to overcome the problem of delaying the selection of a split node to identify the best node and implemented Extremely Fast Decision Tree using Hoeffding Tree to achieve higher Prequential Accuracy than VFDT [29].

Some researchers used the ensemble approach by combining more than one classifier to attain more accurate prediction. Even though it consumes large memory they improved accuracy of the prediction by combining more than one classifier using ensemble approach. Nowadays there is a growing realization of ensemble approaches for streaming data to increase prediction accuracy. Most of the researchers suggested the combination of multiple learners can be more efficient than using a single learner as the outcome is obtained by incorporating the suggestions from the multiple experts than intuitively following a single expert [11–18, 32, 34, 37–39, 44].

From the survey and analysis on comparison among various classification algorithms, it is found that Decision Tree or Hoeffding Tree algorithms are more suitable for streaming data which has high interpretation, accurate, less error rate and they

Table 1 Classification algorithms for streaming data

Algorithm/	Author/year/publication	Merits	Limitations	Research approach
Decision/Hoeffding tree-based algorithms				
VFDT [43]	Pedro Domingos 2000: (ACM)	Needs limited memory, time, and example, able to handle missing values using Hoeffding Bound	Unnecessary tree growth, consuming more time in updating statistics, user-defined threshold value for tie breaking, overfitting due to noise, unable to handle concept drift, and suitable only for static distribution	To handle streaming data
CVFDT [9]	Geoff Hulten 2001: (ACM)	Adaptable with concept drift, dynamically adjusts the size of the window	Forgetting mechanism is slightly complicated, Statistics values are stored at every node to monitor the distribution	To be adaptable with concept drift
$VFDT_c$ [35]	Joao Gama 2003 (ACM)	Includes Naïve Bayes at root node to improve accuracy, does not need previous values of attributes, low variance due to NB, deals with heterogeneous data	Does not deal with concept drift, additional overhead due to deriving NB contingency table to store continuous data, unable to handle noise and bias in data stream, more time to calculate probability using Naïve Bayes at the leaf node	To handle heterogeneous data

(continued)

Table 1 (continued)

Algorithm/	Author/year/publication	Merits	Limitations	Research approach
OcVFDT [33]	Chen Li 2009: (ACM)	Reduces cost by using unlabeled data	Not adaptable with concept drift	To solve class imbalance
Enhanced DT [25]	Ruth Ramya 2012: (IJSER)	Includes Naïve Bayes at the inner node, only stores mean and variance of each stream	Unnecessary tree growth since pruning is not implemented, less accuracy due to overfitting problem of decision tree	To reduce time complexity
Incremental OVFDT [31]	Hang Yang 2013 (Hindawi)	Overfitting, class imbalance problem reduced, deals with noisy data, adaptive tie-breaking threshold is used	Fixing dynamic mean of HB value as tie-breaking threshold value	To overcome overfitting and class imbalance
ECVFT [10]	Gang Liu 2013: IEEE	Examples are cached before execution and no information gain is calculated	Attributes with more information may be lost	To identify the types of drift
stAmine Tree [36]	Mirela Teixeira 2014: JIDM	Information gain and Hoeffding Bound is replaced by statistical mean and standard deviation	Complex to find the statistical values	To overcome the dependency of number of examples

(continued)

are easy to interpret when compared with, Neural Network [40, 41, 45, 46], Nearest neighbor [19, 20, 42, 47], Naive Bayes [20, 21], and Support Vector Machine [22, 30].

Table 1 (continued)

Algorithm/	Author/year/publication	Merits	Limitations	Research approach
VHT [26]	Nicolas Kourtellis 2016	Distributing decision tree through vertical parallelism, using adaptive tie-breaking threshold for node splitting, can process dense and sparse instances	Leaf statistics consistency, tree maintenance, additional hardware cost, memory, communication, and computation complexity between parallel models	To reduce execution time
IPRF [27]	A. Anantha Babu 2017: IJSRST	Time reduction due to parallel implementation	The process is not paralyzed using Spark environment and the examples are not properly spilt through rain forest algorithm	To reduce processing time
VFDT-Hadoop [28]	Sharmishta Desai, 2017: IEEE	Lesser memory and faster using optimal HB value for an efficient tree pruning	Hadoop Map reduce is not a suitable platform for streaming data	To reduce processing time
EFDT [29]	Chaitanya Manapragada 2018: ACM	Learning time is reduced	Computational cost due to revisiting and alternating the best option split	To reduce learning time
OVFDT [23]	Liangde Li 2018: Springer	Naïve Bayes classifier is included at the leaf node to increase accuracy	Less accuracy due to tree-reduction	To reduce tree size and execution time
Strict VFDT [24]	Victor Guilherme 2018: Elsevier	Reduce the unnecessary tree growth and memory usage	As the tree size reduces the predictive performance also reduces	To reduce memory usage and processing time
Ensembled approach algorithms				

(continued)

Table 1 (continued)

Algorithm/	Author/year/publication	Merits	Limitations	Research approach
SEA [44]	Nick Street 2001: ACM	Needs constant memory and adjusts quickly to concept drift. Any time learning method	Accuracy reduces with noisy data. Not suitable for high-speed data streams	To learn stream data at any time
Ensemble [11]	Haixun 2002: ACM	An instance-based pruning, cost-sensitive learning, uses weight-based method for prediction accuracy	Hard to implement	To handle concept drift
MSL [37]	Runxin 2010: KD	A new learner is induced for each concept drift	Error rate increases with less examples	To select the best classifier
DDD [12]	Leandro 2011: IEEE	Improve the prequential accuracy in the presence of drift	Usage of old concept increases memory usage	To handle drift
Ensemble layer [13]	Joao Gama 2014: Springer	Meta learner monitors the learning process, warns or alarms the system if any recurrence of drift occurs	More memory, cost and time consuming	To detect recurrent concept drift
DOED [14]	Parneeta et al. 2015: Springer	Usage of Naïve Bayes and Hoeffding tree as classifiers improves accuracy	Attribute independency is not always possible with all data	To handle concept drift
ASML [38]	Ignasio 2015	Current best model is selected by the deployer	Usage of Multiple regression consumes more cost	To select the best classier

(continued)

Table 1 (continued)

Algorithm/	Author/year/publication	Merits	Limitations	Research approach
FASE [15]	Isvani Frías-2016: ACM	Process the data in constant time and space	Computational complexity due to the usage of perceptron	To handle concept drift
Predict—Detect [16]	Tegjyot 2017: Elsevier	Handled adversarial concept drift, reduces the labeling cost	Difficult to detect adversarial concept drift	To handle concept drift
RVFDT [17]	Dong Zhenjiang 2017: ZTE	Random selection of node splitting attributes	No proper final integration of multiple tree values, significant attributes may be lost due to random selection	To handle concept drift
MLP [34]	Anand Gupta: 2017	Multi-class classifier is combined with multi-class deep learning	Overfitted data set in the first level and more time consumption in the second level	To handle class imbalance
DTEL [39]	Yu Sun (2017)	Using DTEL, traditional learners are preserved for ensemble approach	Memory and computation complexity	To preserve previous model
KME [18]	Siqi Ren 2017: Elsevier	Reduces labeling cost	Less diversity between the base classifiers	To detect multiple drifts
MetaBag [32]	Jihed Khiari 2018	More accurate result due to meta bagging approach	Space complexity due to multiple meta-decision trees and computation complexity	To reduce inductive bias
Neural network-based algorithms				

(continued)

Table 1 (continued)

Algorithm/	Author/year/publication	Merits	Limitations	Research approach
RANMTPR [40]	Tomoyasu 2011: IEEE	Can be used for multiple class data set, one pass incremental learning	Less percentage of interpretation and hard to implement	To learn from multiple class labels
BPTT [45]	Fan He 2017	No need to store all of history parts, Information is stored in each connection of neural network	Usage of Kalman filter's fast convergence, causes overfitting problem	To handle concept drift
LEM [41]	Zeineb Hammami 2017: Elsevier	Dynamic scheduling for resources is performed using regression	Applicable only in offline mode	To extract label from the current model
SASDC [46]	Shizhuo Deng 2017: IEEE	Faster due to parallel distribution	Feedback mechanism consumes more time More computation complexity	Parallel distribution of data set
K-nearest neighbours				
ANNCAD [19]	Yan-Nei Law 2005: Springer	Less time complexity	Difficult to find K value and less accuracy	Adaptable to the concept drift
Hetrostreams [42]	Jesmin 2016: ACM	Euclidian distance and majority votes, reduce the cost of search	More computational complexity	To find the k value
MCNN [47]	Mark Tennant 2016: Elsevier	Micro cluster-based inherently parallel adaptive classifier, calculates statistical summary	Difficult to implement	Parallel implementation

(continued)

Table 1 (continued)

Algorithm/	Author/year/publication	Merits	Limitations	Research approach
SAM [20]	Viktor 2017: Spring	Used self-adjustable memory, ADWIN drift detector	Hard to implement due to self-adjustable memory	To identify Multiple drifts
Naïve Bayes				
OHNBC [20]	Cesar A. 2016: Springer	More accurate due to Naive Bayes approach	Assumptions with respect to the independence of the attributes are not always possible with all data	Histogram-based classification
Naïve Bayes [21]	Sonia 2017: REVIST	Higher accuracy even in 15% noise data	–	To compare various DDM
Support vector machine algorithms				
ILMSSVM [22]	Isaha.: 2017: SACJ	An incremental learning model using SVM, applied incremental k fold cross-validation	–	On line adaptive model
SVM on Haddop [30]	Pranamita Nanda 2017: IRJET	Reduce the time and code complexity, less memory	No proper implementation of Hadoop	To select the most appropriate induction model

5 Performance Metrics

A performance metrics is required to evaluate the quality of any proposed model to implement in real-world problems. Most of the researchers evaluated the following metrics to compare their proposed algorithm with the existing algorithm and demonstrated the efficiency of the proposed classifier.

5.1 Prediction Accuracy

Accuracy is one of the metrics for valuating a classification model. In general, accuracy can be measured by Number of correct predictions/Total number of predictions.

Evaluating the accuracy of a classification model could be a more complex and time-consuming process. Accuracy of a classifier can also be calculated using confusion matrix by finding True Positive True Negative, False Positive, and False Negative values [14, 16–19, 23, 24, 26, 31, 33, 36, 39, 41, 43–45, 47–49].

$$\text{Classification accuracy} = \text{Correct predictions} / \text{Total predictions} * 100.0$$

5.2 *Kappa*

Kappa is similar to Accuracy score, but it measures the kappa value using random predictions. Measuring Accuracy for the algorithm, trained with class Imbalance data set does not provide better accuracy for the less number of classes. In those conditions, finding kappa value improves the quality of any classifier [24, 36].

$$\text{Kappa} = (\text{Observed Accuracy} - \text{Expected Accuracy})/1 - \text{Expected Accuracy})$$

5.3 *Learning Time or Execution Time*

Training Time or Learning Time or execution is Time is time taken to Learn (or) Train (or) Execute by the classifier which is an important factor that affects the suitability of a classification algorithm regarding an unknown data set. An algorithm with a slightly lower accuracy may be preferred if its training time is significantly lower. Moreover, this evaluation is more useful to get the predicted output in a given amount of time [10, 11, 23, 26, 31, 33, 42, 46].

5.4 *Error Rate*

Error Rate is the percentage of incorrect predictions to the number of predictions made, expressed as a value between 0 and 1 [9, 11, 13, 25, 27, 29, 35, 37, 40].

$$\text{Error Rate} = \text{incorrect predictions}/\text{total predictions}$$

5.5 *RMSE (Root Mean Square Error)*

RMSE is another performance measure, especially used for regression problems. RMSE measures the standard deviation of the predictions from the ground truth. This is the relationship between RMSE and classification. A small RMSE denotes a good prediction model and large RMSE denotes the model to be improve [32].

5.6 *Tree Size/Number of Nodes*

Tree size is specifically used to measure the quality of decision tree. Tree size can be measured by the number of nodes used. In general, pruning is used to reduce the tree size. By reducing the tree size, it will lose the generated decision rules and hence prediction accuracy will be reduced [17, 23, 28, 31, 35, 43].

5.7 *Other Metrics*

Apart from the above evaluation metrics, some researchers also used Memory Usage [14, 24], Recall rate [16], and Scalability [42].

6 Streaming Data Platforms (RQ2)

In this section, various streaming data platforms used by the researchers are surveyed and analyzed. Table 2 summarizes the comparative study of the streaming data platforms used. It is found that majority of the researchers [10, 18, 20, 21, 24, 29, 31, 36, 47] used MOA (Massive Online Analytics) framework, due to its ease of use, it also has artificial data generators and amenities to compare various existing algorithm with the proposed algorithm. A. Anadha Babu et al. [27] and Anad Gupta et al. [34] used SPARK platform and Nicoles Kourtellis et al. [26] and Liangde Li et al. [23] used STORM platforms for the distributed and parallel execution to reduce the execution time. Moreover, it is found the usage of Hadoop Map reduce [30, 43] is not suitable since it does not support streaming data as it slows down the process.

7 Conclusion

In this survey paper, we have instigated some of the open challenges and research issues of streaming data. The proportion of research issues solved by various

Table 2 A comparison of open source tools for processing real-time streaming data

	Hadoop Mapreduce	Spark	Storm	MOA
Architecture	Master slave	Master slave	Peer	–
Data processing engine	Supports only batch processing	Supports both batch and stream processing	Supports stream process	Supports stream process
Processing speed	Much slower than others due to map and reduce task	100 times faster than Hadoop due to In memory process	Can process tens of thousands of messages per second on cluster	Slower compared with others
Intermediate storage	Intermediate result is stored in memory	Intermediate result is stored in RDD	No intermediate storage	–
Programming language used	Java, C#, Python and R	Java, Scala, Python, and R	Java and Scala	Java
Scheduler	It needs external Job scheduler like oozie	It has its own scheduler	It has its own scheduler	–
Response time	Minutes to hours	Seconds	Milli seconds	Slower than other
Development cost	Coding available only for batch processing	We can use same coding for stream processing as well as batch processing	We cannot use the same code for both processing	A free Open Source Project
Installation	Easy to install	Easy to install	Not easy to install	Easy to install
References	[30, 43]	[27, 34]	[23, 26]	[10, 18, 20, 21, 24, 29, 31, 36, 47]

researchers are exposed through graphical representation. From the survey, we revealed that the class imbalance and data overfitting issues are addressed by only few researchers which involve a significant role in the prediction accuracy. Then we extended the survey with the existing Classification algorithms of streaming data by analyzing the merits and limitations, which provides a road map for the data scientists and researchers to develop a new algorithm. The algorithm analysis is followed by some of the performance metrics, which are required to evaluate the quality of any proposed model. Finally, we have summarized the comparative study of streaming data platforms used by the researchers and we resolved Hadoop Map Reduce is not suitable for streaming data and most of the implementations are done using Massive Online Analytics (MOA). As a final point, it is concluded that Streaming Data Analytics is a vast research area which includes not only limited to the above-mentioned challenges and there are lots of possibility for the researchers in handling streaming data which should satisfy the basic requirements such as Boundary storage, Speed, and Concept Drift.

References

1. http://www.worldometers.info/world-population/
2. http://opensourceforu.com/2017/09/open-source-tools-you-can-use-to-handle-big-data
3. https://www.idc.com
4. Adrian A (2013) Big data challenge. J Database Syst IV(3)
5. Kandomi A, Haier M (2014) Beyond the hype big data concepts methods and analytics. Int J Inf Manag. ISSN 0268-4012
6. Acharjya DP, Kauser Ahmed P (2016) A survey on big data analytics: challenges, open research issues and tools. Int J Adv Comput Sci Appl (IJACSA) 7(2)
7. Ramírez-Gallego S (2017) Nearest neighbour classification for high-speed big data streams using Spark. IEEE Trans Syst Man Cybern 47(10)
8. Tennant M, Stahl F, Rana O (2017) Scalable real-time classification of data streams with concept drift. Elsevier. https://doi.org/10.1016/j.future.2017.03.026
9. Hulten G, Spencer L, Domingos P (2001) Mining time changing data streams. ACM
10. Liu G, Cheng H, Qin Z, Liu Q, Liu C (2013) E-CVFDT: an improving CVFDT method for concept drift data stream. IEEE. ISBN 978-1-4799-3051-7/13
11. Wang H, Fan W, Yu PS, Han J (2002) Mining concept-drifting data streams using ensemble classifiers. ACM
12. Minku LL, Yao X (2011) DDD: a new ensemble approach for dealing with concept drift. J Res Comput Intell Appl (CERCIA) (IEEE)
13. Gama J (2014) A survey on concept drift adaptations, vol 46, issue 4, article no 44. ACM, Apr 2014
14. Sidhu P, Bhatia MPS (2015) An online ensembles approach for handling concept drift in data streams: diversified online ensembles detection. Springer. https://doi.org/10.1007/s13042-015-0366-1
15. Frías-Blanco I, Verdecia-Cabrera A (2016) Fast adaptive stacking of ensembles. ACM. ISBN 978-1-4503-3739-7/16/04
16. Sethi TS, Kantardzic M (2017) Handling adversarial concept drift in streaming data. Expert Syst Appl
17. Dong Z, Luo S, Wen T, Zhang F, Li L (2017) Random forest based very fast decision tree, vol 15, Dec 2017
18. Ren S, Liao B (2018) Knowledge maximize ensemble algorithm for different types of drift
19. Law Y-N, Zaniolo C (2005) An adaptive nearest neighbor classification algorithm for data stream. In: PKDD, LNAI, vol 3721. Springer, Berlin, pp 108–120
20. Losing V, Hammer B, Wersing H (2017) Tackling heterogeneous concept drift with the self-adjusting memory (SAM). Springer
21. Astudillo CA, Gonzlez JI, John Oommen B, Yazidi A (2016) Concept drift detection using online histogram-based Bayesian classifiers. Springer
22. Jaramillo-Valbuena S (2017) Performance evaluation of concept drift detection techniques in the presence of noise. In: REVISTA
23. Li L, Li P, Xu H, Chen F (2018) A Bayes classifier-based OVFDT algorithm for massive stream data mining on big data platform. Advances in intelligent systems and computing, vol 611. Springer International Publishing. https://doi.org/10.1007/978-3-319-61566-0_49

24. da Costa VGT, de Leon Ferreira de Carvalhob ACP, Juniora SB (2018) Strict very fast decision tree: a memory conservative algorithm for data stream mining. Elsevier. arXiv:1805.06368v2 [cs.A1]
25. Ruth Ramya K, Vishnu Priya RSS, Panini Sai P, Chandrasekha N (2012) Improved decision tree algorithm for data streams with concept-drift adaptation. Int J Sci Eng Res 3(4)
26. Kourtellis N, De Francisci Morales G, Bifet A, Murdopo A (2016) VHT: vertical Hoeffding tree. arXiv:1607.08325V1 [cs.DC]
27. Anantha Babu A, Preethi J (2017) Incremental-parallel data stream classification in Apache Spark environment. Int J Sci Res Sci Technol ICASCT (3)5:201–209
28. Desai S, Roy S, Patel B, Samruddhi P (2017) Very fast decision tree (VFDT) algorithm on Hadoop. IEEE
29. Manapragada C, Geoffrey IW, Salehi M (2018) Extremely fast decision trees. arXiv:1802.0878ov1 [Cs. LG]
30. Lawal IA, Abdulkarim SA (2017) Adaptive SVM for data stream classification. SACJ. ISSN 2313-7835 9
31. Yang H, Fong S (2013) Incremental optimization mechanism for constructing a decision tree in data stream mining. Mathematical problems in engineering, vol 2013, article ID580397. Hindawi Publishing Corporation
32. Khiari J, Matias LM, Shaker A (2018) Meta Bags: bagged meta-decision trees for regression ACM. ISBN 123-4567-24-567/08/06
33. Li C, Zhang Y, Li X (2009) OcVFDT: one-class very fast decision tree for one-class classification of data streams. ACM. ISBN 978-1-60558-668-7
34. Gupta A, Thakur HK (2017) A big data analysis framework using Apache Spark and deep learning. arXiv:1711.09279v1 [cs.dB]
35. Gama J, Fernandes R, Rocha R (2003) Accurate decision trees for mining high-speed data streams. ACM. ISBN 1-58113-737-0/03/0008
36. Cazzolato MT, Ribeiro MX (2014) Classifying high-speed data streams using statistical decision trees. J Inf Data Manag 5(1):84–93
37. Wang R, Shi L, Foghlu MO, Robson E (2010) A META-learning method for concept drift. In: Proceedings of the international conference on knowledge discovery and information retrieval, pp 257–262. https://doi.org/10.5220/0003095502570262
38. Cano I, Khan MR (2015) ASML: automatic streaming machine learning. https://github.com/nachocano/asml
39. Sun Y, Tang K, Zhu Z, Yao X (2018) Concept drift adaptation by exploiting historical knowledge. IEEE Trans 29(10)
40. Takata T, Ozawa S (2011) A neural network model for learning data stream with multiple class labels IEEE Computer Society. ISBN 78-0-7695-4607-0/11
41. Hammami Z, Mouelhi W, Said LB (2017) On line self adaptive framework for tailoring a neural agent learning model addressing dynamic real time scheduling problems. The Society of Manufacturing Engineers, Elsevier. ISSN 0278-6125
42. Tithi JJ (2016) Enabling K-nearest neighbor algorithm using a heterogeneous streaming library: hStreams. ACM. https://doi.org/10.475/123
43. Domingos P, Hulten G (2000) Mining high speed data streams. ACM. ISBN 1-58113-233-6
44. Nick Street W, Kim YS (2001) A streaming ensemble algorithm (SEA) for large-scale classification. ACM
45. He F (2017) Real-time process modelling based on big data stream learning
46. Deng S, Wang B, Huang S, Yue C, Zhou J, Wang G (2017) Self-adaptive framework for efficient stream data classification on storm. IEEE Trans Syst Man Cybern: Syst. ISSN 2168-2216
47. Tennant M, Stahl F, Rana O (2017) Scalable real time classification of streaming data, vol 75, pp 187–199. ISSN 0167-739
48. Domingos P, Hulten G (2014) A general framework for mining massive data streams. J Comput Graph Stat 12(4):945–949
49. Gama J (2010) Knowledge discovery from data streams

Modification of ElGamal Cryptosystem into Data Encryption and Signature Generation

Prerna Mohit and G. P. Biswas

Abstract In the year 1985, Taher ElGamal proposed a new system for encryption and named it as ElGamal public-key cryptosystem. This is considered as one of the most useful security techniques in cryptography. Its security depends on the Diffie–Hellman problem, which internally depends on the discrete logarithm problem. This paper presents an alternate implementation of ElGamal public-key encryption, where a two-tuple ciphertext similar to ElGamal encryption is generated with a difference of using a session key instead of the Diffie–Hellman key to extract the message. Likewise, the ciphertext tuple would be used as a digital signature. Moreover, no security loophole is found in the proposed modification. In addition, the proposed modification has been implemented over the Gaussian integer $Z[i]$ and found to be secured. Finally, it has been claimed that the proposed modification is equally applicable in all security protections where the original ElGamal cryptosystem is used.

Keywords Diffie–Hellman protocol · ElGamal encryption · ElGamal signature · Gaussian integer

1 Introduction

With the rapid development of Internet technology, the security aspects and protection of different Internet-based applications become vital. For instance, the Internet is used in our daily life for banking, purchasing, business, etc. and if the Internet is not well protected and/or security protection is provided in different applications, one is endangering itself by putting credit card numbers and other sensitive/personal

P. Mohit (✉)
Indian Institute of Information Technology Senapati Manipur, Imphal, India
e-mail: prerna215@gmail.com

G. P. Biswas
IIT(ISM), Dhanbad, Jharkhand, India
e-mail: gpbiswas@gmail.com

R. Patgiri et al. (eds.), *Proceedings of International Conference on Big Data, Machine Learning and Applications*, Lecture Notes in Networks and Systems 180,
https://doi.org/10.1007/978-981-33-4788-5_10

information on the Internet. As it is known, the Internet and different layers of the TCP/IP protocol stack are protected by various cryptographic techniques including the IP Secured for the network layer. On the other hand, several data confidentiality, integrity and authentication methods have been developed for the secure transmission of important data over the Internet. Two major approaches namely symmetric-key and public-key encryption are used for data confidentiality operations. However, the latter technique although requires high processing time has more advantages than the former one. Also instead of symmetric, the public-key cryptosystem is mainly used in data authentication, integrity, digital signature, etc. [11–13]. This paper modifies a well-known public-key cryptosystem called ElGamal encryption and proposes an alternative implementation of the same with an advantage that the proposed ElGamal encryption method can be used directly for the generation of digital signatures. ElGamal cryptosystem, which was developed by T. ElGamal in 1985 [1], is an extension of the Diffie–Hellman key exchange protocol, where a similar mask as used in key negotiation is provided in data hiding operations. In 2001, an extended ElGamal cryptography [2] was proposed in the domain of Gaussian integers; however, it was actually implemented using real numbers. After successful implementation of two other public-key cryptosystems called RSA and Rabin in the domain of Gaussian integers [3], ElGamal cryptosystem designed in [2] was corrected and extended over the complex numbers in the domain of Gaussian integers [4]. However, it neither discusses any advantage over the original ElGamal encryption nor presents any security analysis. Also, the ElGamal signature scheme over the domain of Gaussian integers is presented in [5]. An ElGamal-like cryptosystem for the encryption of large messages was proposed by Hwang and Chang in 2002 [6]. Although it is a modification of the existing ElGamal encryption, this scheme has several drawbacks as pointed out by Wang [7]. For example, the security of the Hwang scheme is weaker than the original ElGamal cryptosystem; Wang proved that there are some cases where decryption of the encrypted message leads to no solution. At the same time, Wang also suggested a solution to the above problem as pointed out in [6], but was not able to point out all the problems which made his scheme vulnerable like the use of an XOR operation in a group [8, 9]. After a few years in 2012, Chang [9] proved that the Wang scheme is not sustainable against the chosen-plaintext attack and proposed a new scheme which is sustainable against the chosen-plaintext attack. In 2010, Wei [10] proposed a new characteristic of ElGamal cryptography, i.e. ElGamal encryption is commutative in nature and they define commutative encryption as a process for encryption of a message with two different keys and the decryption operation will always lead to the correct message irrespective of the order of encryption. In this paper, a new approach for traditional ElGamal is specified, which shows that the existing system can be implemented using a randomly generated session key to encrypt a message along with a DH key. Similarly, a digital signature is provided to check the authenticity of the message. In addition, two applications of the proposed scheme (1) Extension over Gaussian integer and (2) Commutative Encryption are also shown. The rest of the paper is organized as follows. It begins with a preliminary Sect. 2 having a brief description of classical ElGamal public-key encryption, Gaussian integer and some of the existing schemes.

Section 3 proposes our modified ElGamal encryption and digital signature as well. Section 4 discusses the application of the proposed modified ElGamal cryptosystem over the domain of Gaussian integers as well as commutative encryption. Section 5 gives the performance analysis and comparison. Finally, the conclusion part is given in Sect. 6.

2 Preliminary

In this section, some relevant topics are discussed, which are related to our proposed scheme.

2.1 *ElGamal Cryptography*

T. ElGamal modified the Diffie–Hellman protocol and presented it in the form of public-key cryptography. The problem of recovering a message given (p, g, g^r, G_p^*) is equivalent to solving the DH problem. In fact, the ElGamal encryption scheme determines the DH session key and multiplies it with the message. Therefore, the securities are considered equivalent.

The ElGamal encryption scheme is described in the setting of the multiplicative group, where p is a large prime number. The multiplicative group of a finite field is cyclic, if and only if it has an element whose order is equal to the order of the group $(p-1)$. That is, there exists an element called primitive root g of a group that can generate all the elements of the group and satisfies $g^{p-1} \equiv 1 mod p$.

Suppose two parties Alice and Bob want to establish a secure connection and perform secret communication over the insecure network. They perform three basic steps (key generation, encryption and decryption) as describe below.

Key generation: Bob selects a finite cyclic group and its generator g and computes $Y = g^d mod p$ where the secrete key is $d \in Z_p^*$. The public key are $\{p, g, Y\}$

Encryption: Alice chooses a secret key $r \in Z_p^*$ and calculates two ciphertexts as

$C_1 = g^r mod\ p$, and

$C_2 = M \times y^r mod\ p$

Decryption: Bob computes the message using ciphertext as

$M = C_2 \times (C_1^d)^{-1} mod\ p.$

The security of the ciphertext depends on both the plaintext and the session key of Alice; as both are unknown parameters, while the security of the plaintext message depends on the private key of Bob, the security of the system depends on the session key of Alice and the private key of Bob as all the other parameters are publicly known.

2.2 ElGamal Digital Signature

As in the real world, the verification of documents can be achieved by a signature in the document, similarly, for the digital world a digital signature is used. ElGamal proposed one of the methods for performing the same as an extension of ElGamal cryptography using the discrete logarithm problem, as there is no feasible solution to this problem. Similar to a cryptographical system, the signature generation method involves three steps: (1) Key generation, (2) Signature of message and (3) Verification of message. The detail is given below.

Key generation: Alice randomly selects secrete key x with $1 < x < p - 1$ and computes $y = g^x modp$.

Signature: To sign a message m, user Alice randomly chooses integer k such that $1 < k < p - 1$ and $gcd(k, p - 1) = 1$.

Computes two signature S_1, S_2 as

$$S_1 = g^k modp$$
$$S_2 = k^{-1}(m - xS_1) modp - 1$$

Verification: user Bob verifies the signature (S_1, S_2) by computing

$$V_1 = g^m mod\ \text{p}$$
$$V_2 = Y^{S_1} \times S_1^{S_2} mod\ \text{p}$$

The verifier accepts a signature if $V_1 = V_2$ is satisfied, otherwise rejects it.

2.3 Gaussian Integer

The Gaussian integer $Z[i]$ is a complex number defined as $(a + bi)$ where a, b are real numbers and i is an imaginary unit: $Z[i] = [a + bi\ |a, b \in Z, i = \sqrt{-1}]$. A complex number with real part zero is purely imaginary, whereas the complex number with an imaginary part zero is a real number. Real numbers are a subset of complex numbers. The norm of a Gaussian integer $\beta = a + bi$ is denoted by $N(\beta) = \beta \times \overline{\beta} = (a + bi)(a - bi) = a^2 + b^2$ or $|a + bi| = a^2 + b^2$ where $\overline{\beta}$ is a conjugation of β. If β divides α in a group, then it is always true that $N(\beta)$ divide $N(\alpha)$. The norm is multiplicative, $N(\alpha\beta) = N(\alpha)N(\beta)$. The concept of a complex number in cryptography extends into the two-dimensional complex plan. The elements of $Z[i]$ having norm value 1 as a unit. It follows that ± 1 and $\pm i$ are units in $Z[i]$. If γ is the greatest common divisor of α and β, its unit multiples are $\gamma, i\gamma, -i\gamma$.

The prime elements of $Z[i]$ are known as Gaussian primes. The Gaussian primes are symmetric in the entire four quadrants over the real and imaginary axes. A Gaussian integer $\beta = (a + bi)$ is Gaussian prime if it falls in any of the following categories:

1. For a Gaussian integer with both real and imaginary parts is nonzero and its norm $N(\beta)$ is a real prime number.
2. If either real part or imaginary part any one of the value is zero and other is prime of the form 3 (mod4).

For example, 359 or 359i is a prime number in Z[i] as it is of the form 4n + 3 where n = 89. The modulo operation is defined using the Gaussian integer division. Let $\beta = (a + bi)$ and $\alpha = (c + di)$ be any two complex numbers, then

$$\beta \; mod \; \alpha = \beta \; \alpha \; \frac{\beta}{\alpha}$$

The term $\frac{\beta}{\alpha}$ is defined as $\frac{(a+bi)}{(c+di)} \times \frac{(c-di)}{(c-di)} = \lceil \frac{ac+bd}{|\alpha|} \rceil + \lceil \frac{bc-ad}{|\alpha|} \rceil i$

The Euclidean algorithm is defined over Z[i] as it is defined over Z for calculating the gcd of two numbers. Let α and $\beta \in Z[i]$ be nonzero numbers; divide $\frac{\alpha}{\beta}$ and find the divisor and the remainder. Again, repeat the division with the dividend and the reminder until the remainder becomes zero. The last nonzero reminder is divisible by all common divisors of α and β, and this value is a common divisor and the gcd of α and β.

$\alpha = \beta \times q_1 + r_1, N(r_1) < N(\beta)$; q_1 is quotient and r_1 is reminder

$\beta = r_1 \times q_2 + r_2, N(r_2) < N(r_1)$; q_2 is quotient and r_2 is reminder

$r_1 = r_2 \times q_3 + r_3, N(r_3) < N(r_2)$; q_3 is quotient and r_3 is reminder

$\vdots$

$r_k = r_{k+1} \times q_{k+1} + r_{k+2}, N(r_{k+2}) < N(r_{k+1})$; q_{k+2} is quotient and r_{k+2} is reminder.

where $r_{k+2} = 0$ and gcd $(\alpha, \beta) = r_{k+1}$.

In a group Z_p^*, a Gaussian integer β is relatively prime to p. The multiplicative inverse modular of a prime number is donated by β^{-1} mod p and defined by $(\overline{\beta})/|\beta| modp = | \beta |^{-1} \times (\overline{\beta})$ mod p.

3 Modified ElGamal Over Integers

As we know, the ElGamal scheme is basically an advancement of the Diffie–Hellman protocol for the design of the cryptographical system where a message is encrypted using the DH key and a random session key is generated; consequently, ElGamal encryption-decryption, ElGamal authentication, ElGamal digital signature, etc. are implemented based on the same concept. For security, a group with a large prime modulo is used. In the proposed modification, a method is developed to encrypt the message using a randomly generated session key rather than a DH key with the same constraint and security, so that in the existing application of ElGamal for providing integrity, authenticity, confidentiality, etc. it could be executed based on the session key, while in the ElGamal scheme it is based on the DH key. Hence, the proposed modification is an alternate approach for the implementation of ElGamal cryptography as well as the ElGamal digital signature, where the computation cost for encryption/decryption is equivalent to the existing system. It may be noted that the proposed ElGamal digital signature is an extension of ElGamal encryption where signature generation uses a DH key as well as a session key, whereas a traditional ElGamal digital signature does not use a DH key for signature generation much similar to the existing schemes. The proposed ElGamal cryptosystem is described below.

3.1 Proposed Algorithm for Confidentiality

The proposed ElGamal cryptography can be described in three step: key generation, encryption followed by decryption is described below:

- Key Generation
  Select a large prime number p over a cyclic group having generator g.
  Calculate $Y = g^d modp$, where $d \in Z_p^*$ is relative prime to $(p - 1)$.
  The generated public-private key pair are $\{p, g, Y\}$, $\{d\}$ respectively.

- Encryption
  Select a private key $r \in Z_p^*$, and calculate the ciphertexts C_1 and C_2.
  $C_1 = Y^r modp$
  $C_2 = Mg^r modp$ where $M < p$.
  The ciphertext C_1, C_2 are send via a secure channel to the desired recipient.

- Decryption
  Select $e \in Z_p^*$ such that $ed^{-1} mod(p - 1)$.
  $M = C_2(C_1^e)^{-1} modp$.

The process of decryption is done in three steps: first is the selection of e such that $d \times e \equiv 1 mod(p - 1)$ by using the extended Euclidean algorithm.The second step is retrieving g^r from C_1, which is computed with the help of e and C_1 where eth power of C_1 will give g^r [4]. The proof of its correctness is shown in the theorem. The third and last step for decryption is dividing C_2 by the result of the second step, as shown in the above decryption algorithm.

Theorem 1 *The modified ElGamal will give the correct decryption of the ciphertext and provide security against an active attack.*

Proof There is an Adversary A, which has a non-negligible advantage in attacking the algorithm. A intercepts the communication and gets the parameters $Y = g^d$ mod p, $C_1 = Y^r$ mod p and $C_2 = M * g^r$ mod p. From these parameters, the adversary tries to compute the value of r or g^r, which is not possible in polynomial time. However, the desired requester (Bob) can compute it, as Bob has the value of its secret key d. The process of computing message M performed by the recipient is explained as follows:

Randomly select $e \times d \equiv 1 mod(p - 1)$
There exists a positive integer x such that
$e \times d = 1 + x(p - 1)$
Compute $C_1^e = g^{erd}$
$(g^{ed})^r = g^{r(1+x(p-1))} modp$
$= g^r \times g^{xr(p-1)} modp$
$= g^r modp \times g^{xr(p-1)} modp$
From Fermat's theorem $g^{xr(p-1)} modp = 1$
Therefore, $C_1^e = g^r$ and the $M = C_2 * (C_1^e)^{-1}$ mod p.

Hence, the modified ElGamal cryptosystem will always give the correct value of the message to the receiver (Bob) after decrypting the ciphertext. For the sake of clarity of the modified ElGamal cryptography over the integer, a simple numerical example is given below.

Example 1: Assume a large prime $p = 359$ having generator $g = 154$, where the private key for receiver $d = 79$ and compute $Y = g^d modp$= 154^{79} mod 359 = 130.
Encryption
Choose a secret key r = 117 and message m = 78, compute two cipher-texts

$C_1 = Y^r modp$ = 130^{117} mod 359 = 166

$C_2 = m \times g^r modp$ = 78 × 154^{117} mod 359 = 181

Send (C_1, C_2) = 166, 181
Decryption
Compute $e = 79^{-1}$ mod358 = 213 and calculate plaintext message $m = C_2 \times (C_1^e)^{-1} modp$ = 181 × $(166213)^{-1} mod$ 359 = 78.

3.2 *Proposed Algorithm for Signature Generation*

As we know, cryptography provides confidentiality of data, which is one of the key aspects of security requirements. In addition, authentication also plays a significant role in it. For data security, both confidentiality as well as authenticity should be applied. We propose a method for digital signature which can be seen as an alternative to the original ElGamal signature scheme. The algorithm for the modified ElGamal signature is approximately the same as the traditional ElGamal signature with some changes in the signature generation phase, such as the calculation of two signature (S_1, S_2). The detailed description is described below.

- Key Generation
  Select a large prime number p over cyclic group Z_p^* having generator g
  Calculate $Y = g^d modp$, where $1 < d < p - 1$
- Signature
  To authenticate message, a private key $1 < r < p - 1$ is selected and compute two signatures S_1 and S_2.

  $S_1 = Y^r mod$p = $g^{rd} modp$

  $S_2 = r^{-1} d^{-1} \times (m - dS_1) mod\ p - 1$

  The signatures along with message S_1, S_2, m is sent
- Verification
  To verify signature, computation of two variables namely V_1 and V_2 is performed

  $V_1 = g^m modp$

  $V_2 = Y^{S_1} \times S_1^{S_2} modp - 1$

  The verifier accepts a signature if $V_1 = V_2$ is satisfied, otherwise rejects it.

4 Application of Modified ElGamal Cryptosystem

Here, we extend our scheme over the domain of Gaussian integer and show a few instances with different combinations of elements of Z[i] to support our work, as we have performed in a modified ElGamal cryptosystem. Initially, a prime number is selected $\in Z[i]$ lets name it β; work in the group of the ring of integers modulo a prime, $Z[i]/<\beta>$. A cyclic multiplicative group of order $(|\beta^2|-1) or \varphi(\beta) = (|\beta^2|-1)$, the order is larger than the classical ElGamal cryptosystem because the set of elements of group G^*_β is larger and to cover all the elements of group it takes more numbers of steps, without any extra effort. Note that G_β is a completely residue system modulo β and G^*_β is a reduced residue system modulo β. For a Gaussian prime β, one can find a generator g in G^*_β by randomly choosing an element g in G^*_β and compute $\varphi(\beta) = N(\beta) - l = p_1^{\alpha_1}, p_2^{\alpha_2} \ldots . p_k^{\alpha_k}$ where $p_1, p_2, \ldots . p_k$ are prime factors of $\varphi(\beta)$. Thus g is a generator if

$$g^{\varphi(\beta)/p_j} \neq 1 mod$$

Otherwise, choose another generator g and repeat the process till a valid generator is set up. The whole process of the scheme over Z[i] is the same as the modified ElGamal cryptosystem, but the integers used $\in Z[i]$ and the calculation of multiplicative inverse modulo are different as described above. The process of the algorithm is given as follows:

- Key Generation
  Generate a Gaussian prime number β having generator g.
  Select a private key $d \in G^*_\beta$ such that it is relatively prime to $N(\beta) - 1$
  Calculate the public key $Y = g^d mod \beta$
  The generated public-private key pair are $\{\beta, g, Y\}$ and $\{d\}$

- Encryption
  Select a private key $r \in G^*_\beta$ and calculate the ciphertexts C_1 and C_2.
  $C_1 = Y^r mod \beta$
  $C_2 = M g^r mod \beta$
  Send C_1, C_2
- Decryption
  Select e such that $d \times e \equiv 1 mod(N(\beta) - 1)$.
  $M = C_2 (C_1^e)^{-1} mod \beta$.

In order to compute g^r from C_1, an integer e is chosen during decryption, such that $de \equiv 1 mod(|\beta^2| - 1)$ or it can be written as $de \equiv 1 mod(\varphi(\beta))$ there exist an integer k such that $de = 1 + k(\varphi(\beta))$. From fermat's theorem if $gcd(g, \beta) = 1$ then $g^{\varphi(\beta)} \equiv 1 mod(\beta)$, and for Gaussian prime β, $\varphi(\beta) = |\beta^2| - 1$, so $g^{|\beta^2|-1} \equiv 1(mod \beta)$.

Thus, $C_1^e = g^{edr} mod \beta = g^{r(ed)} mod \beta = g^{r(1+k(\varphi(\beta)))} mod \beta = g^r * g^{k(\varphi(\beta))} mod \beta = g^r mod \beta$.

From Fermat's theorem, $M = C_2 (C_1^e)^{-1} mod \beta = M g^r \times (g^r)^{-1} mod \beta = M$.

Here, we show three cases with different entities to be an element of Z[i] at different times and prove that in all cases, our scheme is working fine.

Example 2: prime β, generator and message—all the parameters are complex numbers.

Key Generation
Let Bob select $\beta = 17 + 10i$, $g = (2 + 7i)$ and compute $Y = g^d mod\beta = (2 + 7i)^{47}$ $mod(17 + 10i)$ =11i where d is a secret key of Bob.
Bob’s public key is = (17+10i), g = (2+7i), Y=11i while Bob’s private key d= 47.
Encryption
Let Alice chooses the plaintext message m= 7+8i, and calculate the ciphertext using the secret key r= 28 as.
$C_1 = Y^r mod\beta = (11i)^{28} mod(17 + 10i) = (1 + 5i)$
$C_2 = m \times g^r mod\beta = (7 + 8i)(2 + 7i)^{28} mod(17 + 10i) = (8 + 2i)$
Send C_1, C_2 to Bob.
Decryption
Bob selects $e = 47^{-1} mod388$ = 355 and calculates plaintext message $m = C_2 \times (C_1^{e)-1})mod\beta = (8 + 2i) \times ((1 + 5i)^{355)-1} mod(17 + 10i) = (8 + 2i) \times (-7 + 6i)^{-1} mod(17 + 10i) = 7 + 8i$.

5 Performance

5.1 Complexity

The computation cost of a modular exponentiation using the squaring and multiply methods is at most $O(t^3)$, where $t = log_2(r)$, and there are two modular exponentiation operations that are involved during the encryption operation having complexity of $O(log_2(r)^3) + O(log_2(r)^3) = 2O(log_2(r)^3)$, thus total of $O(log_2(r)^3)$. Similarly, the cost of decryption is $O(log_2(d)^3)$ as decryption cost involves multiplicative inverse $(O(log_2 p))$ + modular exponential $(O(log_2(d)^3))$ where $O(log2_p) << O(log_2(d)^3)$. Thus, the total complexity of the encryption + decryption operation is $O(t^3)$, which is equivalent to the existing ElGamal cryptosystem as both involve only one exponentiation during the calculation of complexity.

5.2 Comparison

In this subsection, the comparison of our scheme with some of the existing schemes is shown in Table 1 based on ElGamal cryptosystem in terms of important parameters such as encryption/decryption of message and XOR operator. As was proved, the XOR operation makes a group operation vulnerable to various types of attacks. However, the proposed modification uses modular exponentiations and multiplication operation, which is suitable for group operations

Table 1 Comparison of our scheme with some of the other schemes

Parameters	XOR operation	Encryption and decryption
Hwang scheme	Used, that might	Decryption might
	cause system failure	be unsuccessful
Wang scheme	Used, that might	Modification of Hwang
	cause system failure	(and XOR-based construct)
Chang	Not used	Uses hash function, thus,
		vulnerable to random oracle attack
Our scheme	Not used	Secure and underlying
		construct as in ElGamal scheme

6 Conclusion

In the above work, a new public-key encryption and signature are proposed based on the assumption that the discrete logarithm is a hard problem, whereas the basic model is the Diffie–Hellman key exchange. The modification in existing ElGamal with some change in the process of encryption. Moreover, a random session key is used for the encryption, and provides a more practical cryptosystem than the existing one. It is also extended over Gaussian integer Z[i] and shows that commutative encryption exists in the modified version, whereas other applications which are supported by ElGamal PKC can be implemented using the new public-key encryption technology. In addition, comparisons of our scheme with different existing schemes are shown in terms of different parameters which demonstrates better performance of our scheme.

References

1. ElGamal Taher (1985) A public key cryptosystem and a signature scheme based on discrete logarithms. IEEE Trans Inf Theory 31(4):469–472
2. El-Kassar AN et al (2001) El-Gamal public key cryptosystem in the domain of Gaussian integers. Int J Appl Math 7(4):405–412
3. Elkamchouchi H, Elshenawy K, Shaban H (2002) Extended RSA cryptosystem and digital signature schemes in the domain of Gaussian integers. In: The 8th international conference on communication systems, 2002. ICCS 2002, vol 1. IEEE
4. Winton, R (2007) Enhancing the Massey-Omura cryptosystem. J Math Sci Math Educ 7(1)
5. El-Kassar AN, Haraty RA (2005) ElGamal Public-Key cryptosystem in multiplicative groups of quotient rings of polynomials over finite fields. Comput Sci Inf Syst 2(1):63–77
6. Haraty RA, El-Kassar AN, Shebaro BM (2006) A comparative study of ElGamal based digital signature algorithms. J Comput Methods Sci Eng 6(5, 6 Supplement 1):147–156
7. Hwang Min-Shiang, Chang Chin-Chen, Hwang Kuo-Feng (2002) An ElGamal-like cryptosystem for enciphering large messages. IEEE Trans Knowl Data Eng 14(2):445–446

8. Wang M-N et al (2006) Cryptanalysis on an ElGamal-like cryptosystem for encrypting large messages. In: Proceedings of the 6th WSEAS international conference on applied informatics and communications. World Scientific and Engineering Academy and Society (WSEAS), 2006
9. Chang T-Y, Hwang M-S, Yang W-P (2012) Cryptanalysis on an improved version of ElGamal-like public-key encryption scheme for encrypting large messages. Informatica 23(4):537–562
10. Dai W (2010) Commutative-like encryption: a new characterization of ElGamal. arXiv:1011.3718
11. Mohit P, Biswas GP (2015) Design of ElGamal PKC for encryption of large messages. In: 2015 2nd international conference on computing for sustainable global development (INDIACom). IEEE
12. Mohit P, Biswas GP (2016) Modification of symmetric-key DES into efficient asymmetric-key DES using RSA. In: Proceedings of the second international conference on information and communication technology for competitive strategies. ACM
13. Mohit P, Biswas GP (2017) Modification of traditional RSA into symmetric-RSA cryptosystems. Int J Busin Data Commun Netw (IJBDCN) 13(1):66–73

Preparation of Sentiment tagged Parallel Corpus and Testing Its Effect on Machine Translation

Sainik Kumar Mahata, Amrita Chandra, Dipankar Das, and Sivaji Bandyopadhyay

Abstract In the current work, we explore the enrichment in the machine translation output when the training parallel corpus is augmented with the introduction of sentiment analysis. The paper discusses the preparation of the same sentiment tagged English-Bengali parallel corpus. The preparation of raw parallel corpus, sentiment analysis of the sentences, and the training of a Character-Based Neural Machine Translation model using the same has been discussed extensively in this paper. The output of the translation model has been compared with a baseline translation model using automated metrics such as BLEU and TER as well as manually.

Keywords Machine translation · Sentiment analysis · Parallel corpus · Neural networks

1 Introduction

Machine Translation (MT) is the translation of one language to another with the help of computer software. MT is a difficult task and involves a thorough understanding of the source, as well as the target text [8].

To train a good MT model, a large and good quality Parallel Corpus is required [13], where a parallel corpus is a collection of bilingual translated texts [14]. In simple words, if two languages are involved: the source monolingual text is an exact

S. K. Mahata (✉) · A. Chandra · D. Das · S. Bandyopadhyay
Jadavpur University, Kolkata, India
e-mail: sainik.mahata@gmail.com

A. Chandra
e-mail: amritac03@gmail.com

D. Das
e-mail: dipankar.dipnil2005@gmail.com

S. Bandyopadhyay
e-mail: sivaji_cse_ju@yahoo.com

R. Patgiri et al. (eds.), *Proceedings of International Conference on Big Data, Machine Learning and Applications*, Lecture Notes in Networks and Systems 180,
https://doi.org/10.1007/978-981-33-4788-5_11

translation of the target monolingual text. Unfortunately, these resources are often scarce, limited in size, and have limited language coverage.

Since, MT is a phenomenon by which, at semantic level, the machine translates one language to another, the translation quality takes a hit as state-of-the-art approaches don't dwell into the pragmatic level when translating. Our idea was that of introducing pragmatic features into MT, such that it improves the quality of translation.

Sentiments express the attitude and emotional condition of the speaker. So, it plays a major role in MT [10]. Since, a parallel corpus has a major impact on **S**tatistical **M**achine **T**ranslation (SMT) and **N**eural **M**achine **T**ranslation (NMT), the performance of MT can be increased if sentiment, by any means, can be added to the parallel corpus. For this, we have prepared a sentiment tagged parallel corpus for English-Bengali language pair. Furthermore, we compare the effectiveness of this corpus with a character-based NMT model trained using a non-sentiment tagged parallel corpus, using metrics such as BLEU [11] and **T**ranslation **E**rror **R**ate (TER) [16], and manual evaluation.

Also, it has been established that MT systems work better when trained using simple sentences only [9]. In that event, we wanted to check whether this holds true for sentiment enriched simple sentences as well. Consequently, we prepared two additional sentiments tagged parallel corpora; one comprising of only simple sentences and another consisting of "Other" (complex and compound) sentences.

The paper has been organized as follows. Section 2 describes a brief survey about the work done in this domain so far. Section 3 describes the methodology of text simplification, preparation of the sentiment tagged parallel corpus, and training a character-based NMT model with the same. Section 4 will discuss the results and will show the comparison of the created model with the baseline models. This will be followed by the Conclusion and Future Scope in Sect. 5.

2 Literature Survey

Various works have already been done on text simplification, sentiment analysis, and generation of parallel corpora. Petersen et. al. [12] worked on text simplification for language learners. Simplified texts are commonly used by teachers and students in bilingual education and other language-learning contexts.

Cardie et. al. [3] found out that finding simple, non-recursive, base noun phrases are an important subtask in many **N**atural **L**anguage **P**rocessing (NLP) applications. Denecke [6] introduced a methodology for determining the polarity of text within a multilingual framework. The method leveraged on lexical resources for sentiment analysis available in English SentiWordNet.[1]

Khan et. al. [7] proposed the rule-based domain-independent sentiment analysis method. The proposed method classified subjective and objective sentences from reviews and blog comments. Zanettin [20] worked on how small bilingual corpora

[1] http://sentiwordnet.isti.cnr.it/.

of either general or specialized language can be used to devise a variety of structured and self-centered classroom activities whose aim was to enhance the understanding of the source language text and the ability to produce fluent target language texts.

Bannard et. al. [2] worked on using alignment techniques from phrase-based statistical machine translation, they showed how paraphrases in one language can be identified using a phrase in another language as a pivot. Varga et. al. [18] worked on a general methodology for rapidly collecting, building, and aligning parallel corpora for medium density languages, illustrating their main points on the case of Hungarian, Romanian, and Slovenian.

Resnik et. al. [15] worked on using the STRAND [13, 14] system for mining parallel text on the **W**orld **W**ide **W**eb (WWW). Baccianella et. al. [1] worked on presenting SENTIWORDNET 3.0, a lexical resource explicitly devised for supporting sentiment classification and opinion mining applications.

Pal et. al. [10] worked on how sentiment analysis can improve the translation quality by incorporating the roles of sentiment holders, sentiment expressions, and their corresponding objects and relations.

No work has been done so far, with respect to the creation of a sentiment tagged parallel corpus. On top of that, creating the same corpus with sentences of various complexities have not been explored as well.

3 Methodology

Our method involves the collection of English sentences initially and translating them into their corresponding Bengali counterpart to create a parallel corpus. For the additional resource involving simple sentences, the English data was shallow parsed to get the phrase structures. Using these structures, the sentences were divided into simple and other sentences. Later, sentiment annotation for both the English and Bengali sentences were done, using various lexicons and this led to the preparation of the sentiment tagged parallel corpus. This method is shown in Fig. 1. All the steps are discussed in detail in the following subsections.

3.1 Collection of English Sentences

We had a English-Bengali parallel corpus from **T**echnology **D**evelopment for **I**ndian **L**anguages **P**rogramme (TDIL)[2], which consisted of 49,999 parallel sentence pairs. In addition to this, we collected 57,985 English sentences from the resource of **M**achine **T**ranslation in **I**ndian **L**anguages (MTIL) shared task[3], organized by Amrita University and the statistics are shown in Table 1.

[2]http://tdil.meity.gov.in/.

[3]http://nlp.amrita.edu/mtil_cen/.

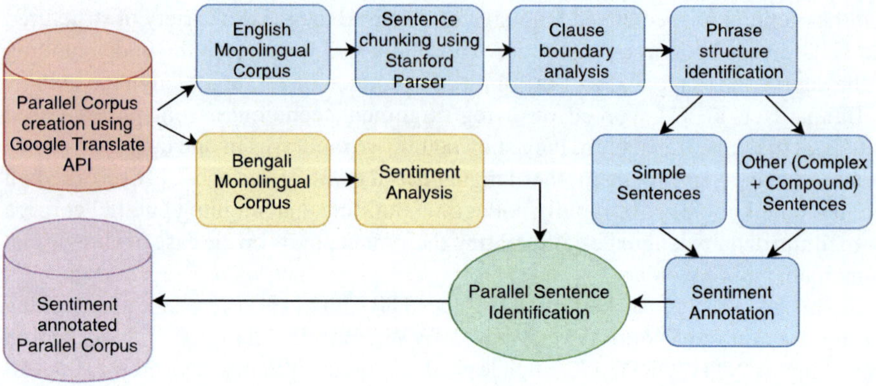

Fig. 1 Methodology

Table 1 Data information table

Source	Data
Amrita University	57,985
TDIL	49,999
Total	1,07,984

3.2 *Translation Using Google Translate API*

The extracted English sentences, apart from the one provided by TDIL, were translated into Bengali using the Google Translate API[4] for Python. Then the English sentences and their Bengali translations were merged to obtain an additional parallel corpus. As a result, the total number of English-Bengali parallel sentence pairs extracted for this experiment was 1,07,984. This corpus has been termed as "General Corpus" for the rest of the paper.

3.3 *Shallow Parsing*

Apart from the preparation of the sentiment tagged parallel corpus, we also wanted to partition the prepared corpus into **I.** simple sentences only and **II.** Other (Complex and Compound) sentences only. This is due to the fact that MT systems work better when trained using simple sentences. So, we decided to create parallel corpora, pertaining to different complexities, enriched with sentiment features as well. To accomplish the classification, shallow parsing was used.

[4]https://pypi.org/project/googletrans/.

Table 2 Example of shallow parsing

Sentence before Parsing
Initially, it ran on 6 routes which joined most of Delhi's parts.
Sentence after Parsing
S (ADVP (RB initially)) (, ,) (NP (PRP it)) (VP (VBD ran) (PP (IN on) (NP (NP (CD 6) (NNS routes)) (SBAR (WHNP (WDT which)) (S (VP (VBD joined) (NP (NP (JJS most)) (PP (IN of) (NP (NP (NNP Delhi) (POS 's)) (NNS parts)))))))))) (. .))

Shallow Parsing is an analysis of a sentence in which constituent parts of sentences (nouns, verbs, adjectives, etc.) are identified and then higher order units that have discrete grammatical meanings (noun groups or phrases, verb groups, etc.) are linked.

We have used Stanford Parser[5] in conjunction with Natural Language Toolkit[6] (NLTK) for performing shallow parsing on the English sentences. We avoided shallow parsing the Bengali sentences as no standard libraries were available for the same. An example of shallow parsing is shown in Table 2.

Hereafter, clause identification of the English sentences was done as discussed in Sect. 3.4.

3.4 *Clause Identification*

3.4.1 Identification of Simple Sentences

A simple sentence is defined as a sentence which contains only one independent clause and has no dependent clauses. Generally, whenever two or more clauses are joined by conjunctions (coordinating and subordinating), it becomes a complex or a compound sentence accordingly. So, to get a hold on handling the conjunctions, we used the Stanford parser to shallow parse the English sentences into phrases. (viz. `NP` (Noun Phrase), `VP` (Verb Phrase), `PP` (Preposition Phrase), `ADJP` (Adjective Phrase) and `ADVP` (Adverb Phrase)).

We noticed that simple sentences have a unique phrase structure that consists of combinations of `NP`, `VP`, and `PP` as shown in Fig. 2. In conjunction with this theory, we applied a rule-based approach to extract simple sentences from the English corpus.

We subjected 3,046 simple sentences to shallow parsing and extracted the unique phrase structures. This constituted the rules by which we further mined for simple sentences from the English corpus. We extracted 205 unique rules, the surface forms of which, along with their confidence score, are shown in Table 3. Confidence Score

[5]https://nlp.stanford.edu/software/lex-parser.shtml.

[6]http://www.nltk.org/.

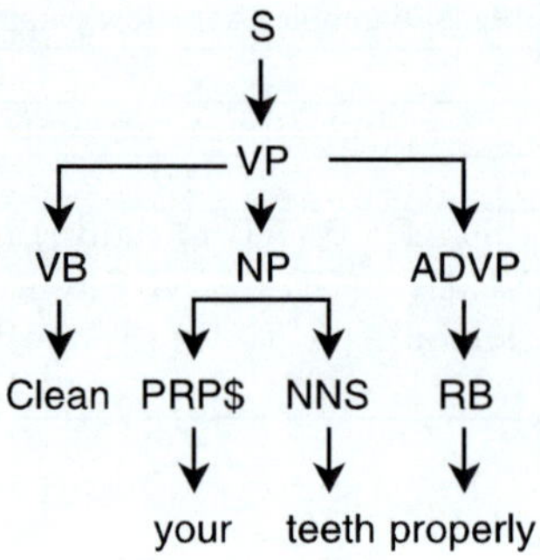

Fig. 2 Extraction of phrase chunks

Table 3 Surface forms of the extracted rules. "*" means one or more occurrence of item

Rules	Confidence
PP NP* PP VP NP*	8.40
PP NP* VP PP NP*	9.49
ADVP NP* VP* ADVP NP*	9.36
NP VP PP NP PP NP	12.15
NP ADVP VP* NP*	11.69
NP* VP NP*	11.69
NP* PP NP VP* NP	11.46
NP VP PP NP*	11.23
VP* NP* PRP* ADVP*	4.92
NP VP* NP* PP* ADJP* ADVP*	9.62

was calculated as a fraction of the total number of sentences identified using a specific rule, by the total number of sentences, as denoted in Eq. 1.

$$\text{Confidence Score} = \frac{\text{No. of sentences identified using rule}}{\text{Total no. of sentences}} \tag{1}$$

We tested our system on 2,876 sentences (1,438 simple sentences and 1438 complex/compound sentences) and got an accuracy of 89.22%. Table 4 shows the confusion matrix for the same. We used this system to extract 16,654 simple sentences from the general corpus.

3.4.2 Identification of Other Sentences

Complex POS tags assigned to every token in the shallow parsed output, were used to identify the complex sentences. Rules followed for extracting complex sentences are as follows, where `SBAR` tag denotes subordinating conjunctions.

Table 4 Confusion matrix for the rule-based approach followed to extract simple sentences

	Other	Simple	Kappa
Other	1275	90	0.78
Simple	220	1291	
Prec.	93.41%		
Recall	85.28%		
Acc.	89.22%		
F1	89.16%		

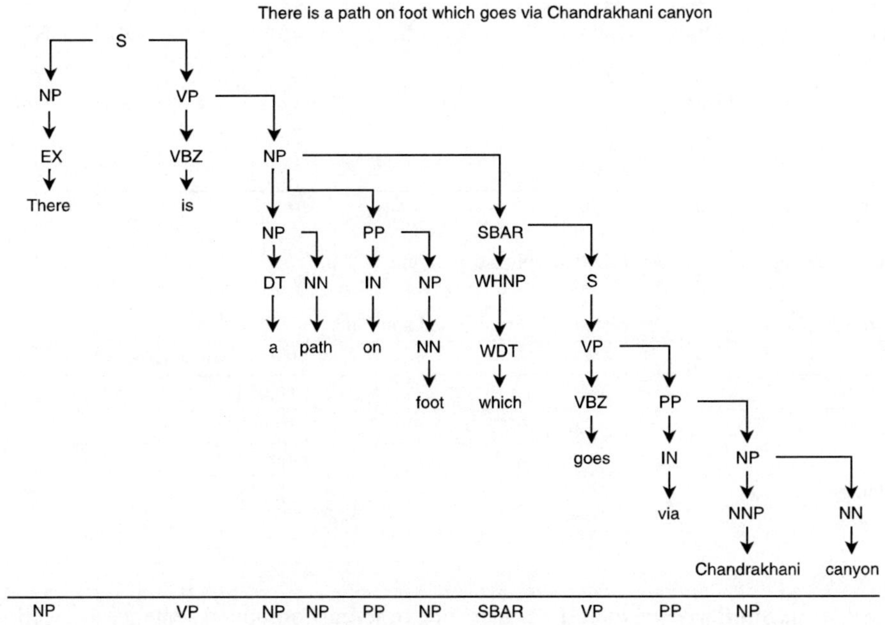

Fig. 3 Example of complex sentence phrase structure

- If a line in a shallow parsed file contains SBAR tag in between two sentences, then we can say that the sentence is a complex sentence.
- If a line in a chunked file contains SBAR in starting of the sentence and ',' in middle of the sentence, then the sentence is considered as a complex sentence.

An example of a phrase structure of a complex sentence is shown in Fig. 3.
Using these rules, we were able to extract 39,068 complex sentences from the general corpus.

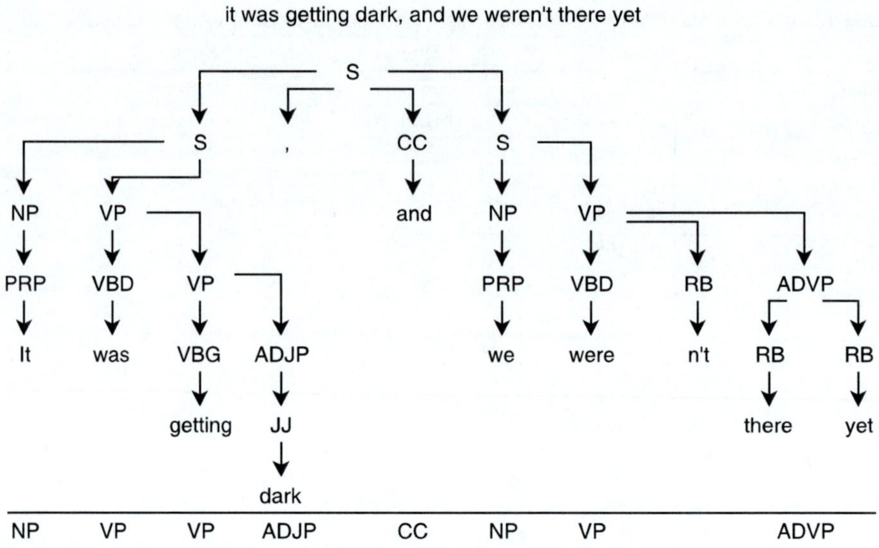

Fig. 4 Example of compound sentence phrase structure

Table 5 No. of sentences extracted using rule-based approach

Type		Number of sentences
Simple		16,654
Others	Complex	39,068
	Compound	43,703
Untagged		8,559

Compound Similarly, we identified rules for extracting compound sentences as well. If the shallow parsed sentence has a CC (coordinating conjunction) tag followed by S (starting of a sentence) tag, then the sentence is a compound sentence. Symbolically, we can define the rule as follows:

$$\mathrm{S}_{Start}\text{...Other POS...}\mathrm{CC}_{POS} \rightarrow \mathrm{S}_{Start}\text{...Other POS}$$

An example of phrase structure of a compound sentence is shown in Fig. 4.
Using these rules, we were able to extract 43,703 complex sentences from the general corpus.

The number of simple, complex, and compound sentences extracted with the help of Sects. 3.4.1 and 3.4.2 are shown in Table 5.

Our system could not tag 8,559 sentences into any category. It is to be noted that, our hypothesis considers that the Bengali sentences have the same complexity, like that of its English counterpart.

3.5 Sentiment Annotation

Sentiment Analysis (SA) refers to the use of NLP to systematically identify, extract, quantify, and study affective states and subjective information. In this work, we have tried to use sentiment annotation in parallel corpus to help increase the performance of MT. This was done by tagging the English words in a sentence S_e and Bengali words in a sentence S_b, with their sentiments, by using SentiWordNets of the respective languages. Here, S_e and S_b are parallel sentences in the general corpus. This process created sentiment tagged English and Bengali sentences S_e^{st} and S_b^{st}, respectively. Consequently, these sentences were termed parallel, if they satisfy certain rules, that has been described in Sect. 3.5.2. Also, sentiment tagging of the English and Bengali words has been described in Sect. 3.5.1.

This experiment was done to study whether enriching parallel corpus with sentiment features enhances the performance of MT or not. Also, since these types of corpora are hard to find, creating one will aid in future research.

3.5.1 Word Level Sentiment Tagging

SentiWordNet of positive and negative words for English[7] and Bengali[8] were applied to the English and Bengali files separately. We also used additional lexicons like AFINN-96[9], AFINN-111[10],Taboada Grieve 2004-SO [17] and Vender Sentiment, for English. This step was repeated for the general corpus, Simple sentence corpus, and the "Other" (Complex, Compound) sentences. It is to be noted that, positive sentiment and negative sentiment were annotated as `POS` and `NEG`, respectively. The neutral sentiments were not annotated. A snippet of the result of this step is shown in Table 6. The result of this step is the generation of sentiment tagged English and Bengali sentences, S_e^{st} and S_b^{st}, respectively.

3.5.2 Sentiment Annotated Parallel Corpus

Sentiment tagged English and Bengali sentences, S_e^{st} and S_b^{st} were considered parallel if they satisfied the following rules.

R1 If the English sentence S_e^{st} is having one or more `POS` tag and its corresponding Bengali sentence S_b^{st} is also having one or more `POS` tag, they are considered as parallel.

---

[7]http://www.nltk.org/howto/sentiwordnet.html.

[8]http://amitavadas.com/sentiwordnet.php.

[9]http://www2.imm.dtu.dk/pubdb/views/publication_details.php?id=5981.

[10]https://udel.instructure.com/courses/1310886/files/57548375.

Table 6 Example of sentiment annotation in both English and Bengali

English sentiment annotated sentence	Bengali sentiment annotated sentence
The tourists admired\POS the paintings.	পর্যটকরা পেইন্টিংসকে প্রশংসিত\POS করেছেন।
I admired\POS him for his honesty.	আমি তার সততা\POS জন্য তাকে প্রশংসিত।
The enemy\NEG soldiers submitted to us.	শত্রু\POS\NEG সৈন্য আমাদের জমা\POS\NEG দেওয়া।
Ellen warned\NEG Helen that the party\POS would be tonight.	এলেন হেলেন কে সতর্ক\POS\NEG করে\NEG দিয়ে বলেন পার্টি আজ রাতে হবে।

Table 7 Statistics of extracted sentiment tagged parallel corpus

Corpus	No. of sentiment tagged parallel sentences
General	70,357
Simple	6,700
Others	63,619

R2 If the English sentence S_e^{st} is having one or more NEG tag and its corresponding Bengali sentence S_b^{st} is also having one or more NEG tag, they are considered as parallel.

R3 If the English sentence S_e^{st} is having one or more POS and NEG tag and its corresponding Bengali sentence S_b^{st} is also having one or more POS and NEG tag, they are considered as parallel.

R4 If the English sentence S_e^{st} is having one or more POS and NEG tag and its corresponding Bengali sentence S_b^{st} is also having one or more POS tag, and vice versa, they are considered as parallel.

R5 If the English sentence S_e^{st} is having one or more POS and NEG tag and its corresponding Bengali sentence S_b^{st} is also having one or more NEG tag, and vice versa, they are considered as parallel.

It is to be noted that, sentences with no POS and NEG tags were not considered as it did not solve the purpose of enriching the sentiment tagged parallel corpus. These steps were followed for the general, simple sentence, and the "Other" sentence corpus. The statistics of the above is given in Table 7.

3.6 Character Based Neural Machine Translation

The Neural Machine Translation model (sequence to sequence model) is a relatively new idea for sequence learning using neural networks. It has gained quite some popularity since it achieved state-of-the-art results in machine translation task. Essentially, the model takes as input a sequence $X = \{x_1, x_2, ..., x_n\}$ and tries to generate the target sequence as output $Y = \{y_1, y_2, ..., y_m\}$, where x_i and y_i are the input and target symbols, respectively.

The architecture of seq2seq model comprises of two parts, the encoder and decoder. Since we wanted to test our model at character level, the inputs to the encoder and decoder were characters of the source and target sentences. We implemented both the models using the Keras [4] library.

Character level NMT (CNMT) performs better than Word Level NMT due to the following reasons:

- It does not suffer from out-of-vocabulary issues
- It is able to model different, rare morphological variants of a word
- It does not require segmentation [5].

Generally, CNMT works best when the majority of alphabets, in the source and target language, overlap, i.e., both the languages share a common or similar script. Still, we tried to find out its performance on the simple sentence and whole corpus, though in our case Devanagari script and Roman script utilizes completely different alphabets. The model has two parts (encoder and decoder) as discussed below.

3.6.1 Encoder

For building the encoder, we used an embedding layer and two **L**ong **S**hort **T**erm **M**emory (LSTM) layers. An embedding layer turns positive integers into dense vectors of fixed size. One hot tensor of the English sentences (embeddings at character level) were fed to the input layer. Thereafter, the embedding layer converted these into dense vectors. From the encoder, the internal states of each cell were preserved and the outputs were discarded. The purpose of this is to preserve the information at the context level. These states were then passed on to the decoder cell as initial states.

3.6.2 Decoder

For building the decoder, again an LSTM layer was used with initial states as the hidden states from the encoder. It was then fed to another LSTM layer. It was designed to return both sequences and states. Here, the concept of teacher forcing [19] was used. The input to the decoder was one hot tensor (embeddings at character level) of Bengali sentences while the target data was identical, but with an offset of one time-step ahead. This was fed to an embedding layer and then to an LSTM layer. The information for generation is gathered from the initial states passed on by the encoder. Thus, the decoder learns to generate target data [t+1,...] given targets [..., t] conditioned on the input sequence. It essentially predicts the output sequence, one character per output time step.

For training the model, various sets of English and Bengali sentences were fed to the CNMT model. These sets will be discussed in Sect. 4. The *batch size* was set to 64, *number of epochs* was set to 100, activation function was *softmax*, optimizer

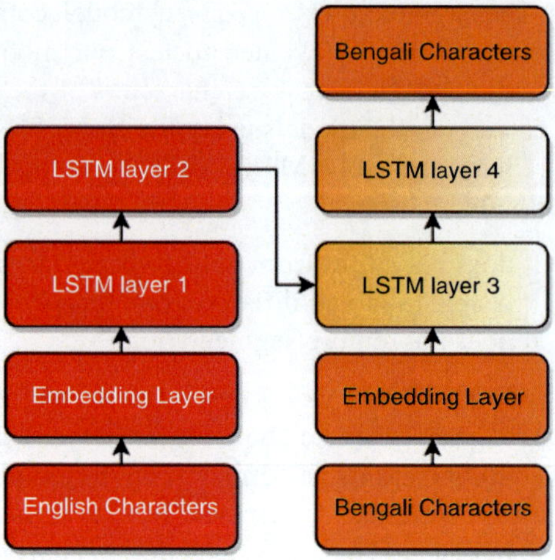

Fig. 5 Architecture of the character embedded NMT model

chosen was *rmsprop* and loss function used was *categorical cross-entropy*. Learning rate was set to 0.001. The general architecture of the created model is shown in Fig. 5.

4 Results

Three sets of experiments were performed with the prepared data. They are listed below.

- Comparison between the general corpus of size 1,07,984 and sentiment tagged general corpus of size 70,357, sentence pairs.
- Comparison between non-sentiment tagged simple sentence corpus of size 16,654 and sentiment tagged simple sentence corpus of size 6,700, sentence pairs.
- Comparison between non-sentiment tagged other sentence corpus of size 82,771 and sentiment tagged other sentence corpus of size 63,619, sentence pairs.

CNMT model was trained using the aforementioned data and automated and manual evaluation of the same is shown in Sect. 4.1.

4.1 Evaluation

To create the CNMT models, that would be used to test the effectiveness of the created sentiment tagged parallel corpus on MT, we have used a non-sentiment tagged parallel corpus as well as the sentiment tagged parallel corpus (General, Simple and Other). All the created models were tested using 200 sentences. The comparison between all the models, with respect to automated evaluation, is shown in Table 8.

Table 8 Automated evaluation of the CNMT model when trained with various datasets

Type	Metric	General	Simple	Other
Non-sentiment tagged	BLEU	10.25	5.67	8.92
	TER	81.01	92.14	89.27
Sentiment tagged	BLEU	11.51	4.68	9.16
	TER	78.59	95.47	87.54

Table 9 Manual evaluation of the CNMT model when trained with various datasets

Type	Metric	General	Simple	Other
Non sentiment tagged	Adequacy	2.98	1.91	2.76
	Fluency	2.56	1.47	2.67
Sentiment tagged	Adequacy	3.03	1.59	2.89
	Fluency	2.84	1.62	2.73

Translation quality was judged by a linguist with Bengali as his mother tongue. The evaluation criteria were Adequacy and Fluency. Adequacy means how much of the meaning is expressed in the target translation. Fluency means to what extent the translation is well-formed grammatically, contains correct spellings and intuitively acceptable, and can be sensibly interpreted by a native speaker. The linguist was asked to rate the translation in the range of 1–5, where "1" is the lowest and "5" is the highest. The manual evaluation measures for the CNMT model, when trained using various datasets, is given in Table 9.

5 Conclusion and Future Scope

In this paper, we have discussed the procedure we followed to build a sentiment tagged parallel corpus. Additionally, we have separately created two resources viz., a sentiment tagged parallel corpus with only simple sentences and another one with "Other" sentences. We can clearly see from the automated and manual evaluation that an NMT model when trained using a sentiment tagged corpus, perform slightly better in both BLEU, and TER, when compared to the baseline system trained using non-sentiment tagged parallel corpus. We can also see from the manual evaluation that fluency improves moderately with the sentiment tagged corpus. Therefore, we can safely say, sentiment does play a role in improving MT output quality.

As a future work, we would like to experiment the same for SMT and Word level NMT. Moreover, since the tagging of sentiments has been done on a word level, using SentiWordNets, we would like to find the sentiment of the entire sentence using a sentiment analysis system and representing the sentiment in the encoder representation. Also, we would like to check whether enriching parallel corpus with sentiment features leads to the propagation of sentiment through the MT pipeline. If yes,this would greatly reduce post editing efforts as well.

Acknowledgements This work is supported by Media Lab Asia, MeitY, Government of India, under the Visvesvaraya Ph.D. Scheme for Electronics & IT.

References

1. Baccianella S, Esuli A, Sebastiani F (2010) Sentiwordnet 3.0: an enhanced lexical resource for sentiment analysis and opinion mining. In: Proceedings of LREC
2. Bannard C, Callison-Burch C (2005) Paraphrasing with bilingual parallel corpora. In: Proceedings of the 43rd annual meeting on association for computational linguistics. Association for Computational Linguistics, pp 597–604
3. Cardie C, Pierce D (1998) Error-driven pruning of treebank grammars for base noun phrase identification. In: Proceedings of the 17th international conference on computational linguistics, vol 1. Association for Computational Linguistics, pp 218–224
4. Chollet F et al (2015) Keras. https://keras.io
5. Chung J, Cho K, Bengio Y (2016) A character-level decoder without explicit segmentation for neural machine translation. CoRR abs/1603.06147. http://arxiv.org/abs/1603.06147
6. Denecke K (2008) Using sentiwordnet for multilingual sentiment analysis. In: IEEE 24th international conference on data engineering workshop, 2008. ICDEW 2008. IEEE, pp 507–512
7. Khan A (2011) Sentiment classification by sentence level semantic orientation using sentiwordnet from online reviews and blogs. Int J Comput Sci Emerg Technol 2(4)
8. Mahata SK, Das D, Bandyopadhyay S (2018) Mtil2017: machine translation using recurrent neural network on statistical machine translation. J Intell Syst, 1–7
9. Mahata SK, Mandal S, Das D, Bandyopadhyay S (2018) Smt vs nmt: a comparison over hindi & bengali simple sentences. arXiv:1812.04898
10. Pal S, Patra BG, Das D, Naskar SK, Bandyopadhyay S, Genabith JV (2014) How sentiment analysis can help machine translation. In: Proceedings of the 11th international conference on natural language processing. NLP Association of India, pp 89–94. http://www.aclweb.org/anthology/W14-5114
11. Papineni K, Roukos S, Ward T, Zhu WJ (2002) Bleu: a method for automatic evaluation of machine translation. In: Proceedings of the 40th annual meeting on association for computational linguistics. Association for Computational Linguistics, pp 311–318
12. Petersen SE, Ostendorf M (2007) Text simplification for language learners: a corpus analysis. In: Workshop on speech and language technology in education
13. Resnik P (1998) Parallel strands: a preliminary investigation into mining the web for bilingual text. In: Third conference of the association for machine translation in the Americas. Springer, pp 72–82
14. Resnik P (1999) Mining the web for bilingual text. In: Proceedings of the 37th annual meeting of the association for computational linguistics, pp 527–534
15. Resnik P, Smith NA (2003) The web as a parallel corpus. Comput Linguist 29(3):349–380
16. Snover M, Dorr B, Schwartz R, Micciulla L, Makhoul J (2006) A study of translation edit rate with targeted human annotation. In: Proceedings of association for machine translation in the Americas, vol 200

17. Taboada M, Grieve J (2004) Analyzing appraisal automatically. In: Proceedings of the AAAI spring symposium on exploring attitude and affect in text: theories and applications, pp 158–161
18. Varga D, Halácsy P, Kornai A, Nagy V, Németh L, Trón V (2007) Parallel corpora for medium density languages. Amst Stud Theory Hist Linguist Sci Ser 4(292), 247
19. Williams RJ, Zipser D (1989) A learning algorithm for continually running fully recurrent neural networks
20. Zanettin F (1998) Bilingual comparable corpora and the training of translators. Meta: Journal des traducteurs/Meta Transl J 43(4):616–630

Privacy-Preserving Association Rule Mining in Distributed Database Environment: A Review

Roberto Jimenez and Luis Ortiz-Ospino

Abstract The research in this paper analyzes the origin of the data generated and treated as Big Data and discusses the hypothesis about the conflict generated by the management of personal data privacy. It is based on a model to identify the set of management practices and resources, resulting from the integration of guidelines, norms, standards and contractual commitments that are more relevant to keeping personal data safe and generating trust in the environment. The field study shows the importance and urgency of this issue, given that it is necessary to innovate in business management, the provision of public services and the design and implementation of regional development policies. Through its conclusions, it is verified that an intelligent management is needed to rectify the course in terms of discovering and detecting patterns, relations and formulating models from these gigantic databases. This work presents a generic model for the management of privacy in the cloud of huge volumes of data. The impact of privacy is analyzed, risks are identified, and solutions provided by standards are explored to develop controls that can be integrated into a model to establish the steps so that any type of public or private organization can verify the organizational impact of its products, processes or service in the fulfillment of its strategic objectives.

Keywords Big Data · Standards · Database and large data management · Data models · Data privacy · Security

1 Introduction

The management and analysis of large data generated by the use of digital technologies (such as mobile communication, electronic transactions and social networks) are

---

R. Jimenez (✉)
Universidad de la Costa, St. 58 #66, Barranquilla, Atlántico, Colombia
e-mail: rjimenez12@cuc.edu.co

L. Ortiz-Ospino
Universidad Simón Bolívar, Barranquilla, Colombia
e-mail: lortiz27@unisimonbolivar.edu.co

R. Patgiri et al. (eds.), *Proceedings of International Conference on Big Data, Machine Learning and Applications*, Lecture Notes in Networks and Systems 180,
https://doi.org/10.1007/978-981-33-4788-5_12

effective tools for innovation in business management, public service delivery and the design and implementation of development policies. This phenomenon known as Big Data [1] represents the set of technologies that allow processing massive amounts of data from disparate sources, with the aim of providing them with a service that provides value. It is a new paradigm, which refers to the enormous growth in the access and use of automated information.

Data collection by corporations and government threatens privacy while promoting transparency [2]. Today more than ever, the growing and enormous amount of data, on the order of zeta bytes, generated by business and e-government applications, put privacy in the cloud at risk and determines the need for an Integration model for security management of these huge data sets.

With the size and complexity of Big Data, a myriad of legal and regulatory challenges arises: Who owns the data? Why do companies use these data? To whom do these huge volumes of data belong? etc. However, the answer is not unique. Ownership of data has many facets depending on how it is generated and treated under strict security measures. Users give away their consumption data and locations mainly for a more comfortable and simple life. Last year's Mobile Congress [3] clearly showed that new business niches are emerging in areas such as health applications, preventive vehicle maintenance, electricity network management, water infrastructure management and intelligent sensor information processing, etc.

Nevertheless, all these guidelines, regulations, laws and contractual commitments become overwhelming, because of the variety and multidisciplinary nature they present, so new business models are needed where privacy management is technically feasible and potentially attractive to the market [4]. This will undoubtedly demand some future models where there should be a dialogue among technical, regulatory, business and civil agents.

This paper analyses these modulations of subjectivity in the digital era, in order to establish common guidelines that allow for an efficient and effective management of data generated by public services, social services and financial services, based on a model that allows for the integration of more relevant standards to keep privacy safe.

2 Big Data

As expressed by [5, 6], the big data begins with the fact that there is a lot of information these days and it is more available than ever for extraordinary uses. Big data is more than just communication: The idea is that one can learn from a large body of information things that one cannot understand when using only small amounts. Although there are several definitions of Big Data, for the purposes of this paper, it will be conceptualized [7] as the huge volumes of data (structured, semistructured and unstructured in the order of exabytes 10–18 bytes), whose storage and analysis, i.e., text analysis of emails, tweets, blogs, can be done through specialized databases

[4] that serve to analyze these huge volumes of data and thus get information and knowledge [8].

As expressed in [9] "... it is, in the field of information and communication technologies, a reference to systems handling large data sets." The most common difficulties in these cases focus on capture, storage, search, sharing, analysis and display. The trend to manipulate huge amounts of data is due to the need, in many cases, to include the related analysis data in a large related data set, such as business analysis, infectious disease data, or the fight against organized crime.

The upper processing limit has shifted over the years, so the limits that were set in 2008 were in the range of petabytes to zettabytes of data. Internet traffic in 2016 was 1.3 ZB, according to a CISCO report [10].

Scientists regularly find limitations due to the large amount of data in certain areas, such as meteorology, genomics, connectomics, complex simulations of physical processes, and research related to biological and environmental processes. Limitations also affect internet search engines, financial systems and business informatics [11].

Data sets are growing in volume due partly to the introduction of ubiquitous information from wireless sensors and mobile devices (e.g., VANETs), the constant growth of historical applications (e.g., blogs), cameras (remote sensing systems), microphones, radio-frequency identification readers. Humanity's data storage capacity has doubled at a rate of forty months since the 1980s [12].

Technologies related to data analysis include large databases such as NOSQL, Hadoop and Map Reduce. The open source Hadoop platform has become synonymous with Big Data for much of the storage industry and analytical applications. Its adoption is not yet massive, but it is progressive.

When talking about Big Data [9], one of the great solutions is to take out all the data that is sitting in Data Warehouses and apply Data Analytics to understand where customers want to go [13]. For example, SIRVE (Sistema Integral de Relevamiento Virtual y Estadístico) [14], developed by Fundación Pescar and the Telecom Group, provides social organizations with a data collection tool that allows them to easily, efficiently and quickly gather information through cell phones and tablets in the same field of work.

Although the Big Data phenomenon is global, it can have an actual and potential economic impact, benefiting both the public and private sectors, on increasing productivity, sectoral competitiveness and the quality of life of citizens [15].

In Argentina, there is an alliance for large data between MINCYT and the Sadowsky Foundation, which presented a strategy for the use of large data for Argentina in the period 2013–2018 [16]. The industries that are making the most progress with the implementation of Big Data are Telco, media, entertainment, services and manufacturing. In the case of Latin America, Argentina, Chile and Peru are making progress, driven by the use cases of other pioneering companies in the region. "The fact that we already have use cases in Latin American countries, starting with Brazil, is allowing us to reduce the adoption and implementation time in other countries, so we are starting to see faster progress." In this sense, Big Data emerges as a new phase in the intensive information and communication paradigm that includes not only its technological dimension but also social, economic, political and cultural dimensions [17].

However, these new opportunities of Big Data are increasing the risks and perhaps the most relevant is the one it represents for the privacy of individuals, with the size and legal complexity also arise a myriad of challenges in legal and regulatory matters [18].

3 Ownership of Data

In its work, the World Economic Forum proposes that although data refer to individuals, it is created from the interaction of various parties, so that all of them should have rights and responsibilities over that information, the rights should be common, not exclusive. To answer the first question of whom the data are, it should be outlined that the owner of the computer resources must be able to prove in some way that he has taken the necessary precautions to ensure the security of his systems in an environment of integrity, confidentiality and availability. Data protection rules [19] ensure that individuals have control over the information that is entered into databases, and the focus should no longer be on the time of data collection, but rather on the time of use of their data.

Regarding the second question about why companies use my data, the Argentine legislation only makes it clear that it is the company that provides the means to elaborate a database, and who has a series of rights over it. This should be spoken of in terms of ownership, or rather, use as established in a certain agreement [20]. Depending on the type of data involved, the access criterion applied is different [21]. The main risk factors include those expressed in Sects. 3.1, 3.2 and 3.3.

3.1 Security Cameras, Surveillance

Surveillance has increased with the proliferation of video surveillance devices, or those that use the human body itself as a support, such as biometrics, and whose use extends to diverse fields such as those used for defense, security and immigration used in government areas [4].

The capture of images in public streets is not limited to the behavior of some citizens who capture images with their cameras, but also includes the use of images captured by private security companies and those belonging to state surveillance, which also adds the enormous number of images generated by the proliferation of drones.

In order to track the movements of buyers and since GPS signals are often not perceived inside sales areas, some retailers may use other electronic technologies. These technologies include RFID on products or shopping carts, video cameras and several innovative technologies taking advantage of mobile phones including facial recognition. In Argentina and the rest of Latin American countries, there is not much awareness about the use of biometric data. These video cameras can be excellent for traffic flow management, but difficult to use for tracking the behavior of individuals who come into conflict with the Data Protection Act, unless the images are distorted to prevent individual identification [22].

Finally, the study mentions the Directive 680/2016 of the European Parliament and of the Council on natural persons responsible for police and judicial matters [23].

3.2 Medical Data

The European Council points out that medical data are part of the individual's privacy, so that they can only be transmitted or disclosed on very specific and restricted issues and problems. In this regard, Article 6 of the Convention on the Protection of Individuals with regard to Automatic Processing of Personal Data states that personal data concerning health cannot be processed automatically unless national legislation provides appropriate security measures. In this sense, Law 25326 provides a safety standard by establishing that health-related data may only be processed for the purpose of offering goods and services when they have been obtained in accordance with the guiding principles established by the law and do not cause discrimination. These data may not be transferred to third parties without the prior, express and informed consent of the owner of the data, who must be given clear notice of the sensitive nature of the data and that he is not obliged to provide them, together with his right to request their removal from the database. The European Union Regulation on Data Protection published on 27/04/2016, established that the consent must be "Free, specific, informed and unequivocal" [24].

As expressed in [15], "All medical care activities aimed at obtaining, classifying, using, administering, guarding and transmitting patient information and documentation must observe strict respect for human dignity and autonomy of will, as well as the due protection of patient privacy and confidentiality of sensitive data in accordance with Law 25326" [25]. The latter is accentuated after the reform of the National Civil and Commercial Code (C.C.C.), especially by the established principles: Good Faith (Art. 9), Abuse of Rights (Art.10), Abuse of Dominant Position (Art.11), Public Order (Art.12) and Prohibition of Renunciation of Laws [26].

In addition, the Law on Patients' Rights, No. 26529, defines informed consent in Article 5 as "the declaration of sufficient willingness made by the patient, or by his legal representatives where appropriate, issued after receiving clear, precise and adequate information from the professional involved" and No. 26742, Decree 1089/2012 and Article 59 of the CCC [27].

Finally, the Good Clinical Practice Guidelines (GCP-ICH) must also be taken into account. The Helsinki Declaration and the provisions of ANMAT N°5330/97, 690/05,1067/08,6550/08 Res BN°1490/07, Law 11044 and Decree 3385/08 and what is established by the Ministry of Health through Decree 3385/08 on informed consent [28].

3.3 Cloud Services

As the authors state, the overall objective of The Cloud is to understand the security, privacy and trust challenges and to advise on policies and other interventions that should be considered to ensure that European users of cloud environments have appropriate protections and to sustain European cloud leadership. Cloud computing is increasingly subject to the interest of policymakers and regulatory authorities. The European Commission's recent Digital Agenda 1 highlighted the need to develop a pan-European "cloud strategy" to support growth and jobs and create an innovation advantage for Europe [29].

However, the concern is that there is currently a number of security, privacy and trust challenges and risks that may undermine the achievement of these broader policy objectives. The authors cited above also analyze the technological, operational and legal complexities of cloud computing, considering the European dimension and the interests and objectives of all stakeholders (citizens, individual users, businesses, cloud service providers, public authorities). Its study represents an evolutionary progression in the understanding of the implications of cloud computing for security, privacy and trust [30].

Leading companies in the industry have recently analyzed the threats and risks of cloud computing and made recommendations. Their concerns are reflected in aspects such as data management, mainly in the ownership of data and the way it is operated and treated by providers, and in the identification and control of access to resources. In addition, it includes security reinforcements in terms of data confidentiality [31].

The Open Cloud Manifesto establishes a set of principles to ensure an open cloud. Open collaboration and appropriate use of standards to meet the challenges of cloud deployment. Service providers must not retain users on certain platforms, preventing freedom of choice. Where appropriate, providers should use existing standards. Any new standards adopted should promote innovation and under no circumstances restrict it [32].

4 Privacy

The Big Data defies all data protection rules by facilitating the reidentification of subjects, not only from pseudonymous data but also from data that we consider anonymous, and can also be personal data, considering the broad concept promoted by the Community rules. Health data that are then anonymized and processed for research purposes, given that Big Data increases the amount and diversity of information, facilitates the reidentification of individuals and even after they have been anonymized [19]. Some models have been proposed to examine the re-identification attacks that can be carried out in versions that adhere to anonymity as long as the accompanying policies are respected [33].

Within this framework, the European Union published the European Union Regulation on Data Protection on 27/04/2016, which, although it is directly applicable in all member states from the moment of its publication, has established that it will be in force from May 2018 [34]. It includes concepts such as privacy by design, which requires that privacy be taken into account from the initial phase, from the very design of the product or service, thus not only achieving greater efficiency in the protection of the rights of those affected. The inclusion of these concepts will oblige companies to update their internal processes to adapt them to these requirements. Furthermore, from the application of this Regulation, the report on the impact of privacy on intimacy will be obligatory.

One of the proposed models, which is being tested in countries such as the UK, is the creation of permits for the processing of different data depending on the context in which they are to be used (e.g., in the health sector, financial sector, etc.) [35].

The concept of fingerprinting has generated countless debates and controversies about what is personally identifiable information. However, a problem has become apparent because personal fingerprints have been found on data that until now had not been considered personally identifiable. That is to say, despite the fact that in a database no names appear, patterns can be seen and with these patterns, a person with sufficient analytical knowledge can obtain names.

Information technologies have generated a new concept of identity: the digital one [36]. The profile of the "virtual person" is defined in the network and is fed by the content that it provides about a particular individual or company: the web provides content, identifies and individualizes the person in one way or another.

In this way, the digital identity represents the same characteristics and activities as the real identity, but carried out on the Internet, as a consequence of the growth of digital communications. This identity is what is referred to as "virtual life."

Some authors conceptualize it as "The set of particular features and characteristics that a person expresses through the Internet forms as an essential part of the personal identity of each subject, in its dynamic aspect, and more precisely in its psychological, social and moral aspect.

The digital identity is actively constructed [4], contributing texts, images and videos to the Internet, participating, definitively, in the web world. On social networking sites, it is built from a user profile, which is often linked to profiles of other users or contacts. Many sociological, cultural and even psychological aspects converge in the digital identity [6].

For [9], it is a representation of an active entity in the computer. This entity can be physical (user, server or another device) or software. The identity is assigned to an identifier which in turn contains attributes and rights, which are referenced to a profile. An identity management system aims to make the creation, assignment of rights or denial of permissions of a profile as simple as possible. Therefore, it is possible to define the identity management as the administration of the same one, under established standards so that the security of the information is the correct one.

5 Model of Privacy

The mechanisms of international legality are necessary because of the risks that Big Data entails for privacy. This has changed the focus of attention; it is no longer necessary to comply only with the principles of data protection. Now, privacy must be taken into account from the same business model, that is to say that privacy is no longer a legal concept but a business priority.

Given that the Data Protection Impact Assessment model is not sufficiently advanced and developed, but is a first approach generated from the referenced project [13, 37], it is in the process of being contracted and is a necessary budget to provide data controllers with sufficiently specific, useful and clear practical guidance.

Organizations develop a set of activities and processes, which must be systematically managed to allow them to meet their objectives [10]. This leads the work team to think that Information Systems (IS) need an interdisciplinary vision and evaluation. In this sense, a new perspective appears defined as IS Artefact that has as components: the IT Artefact, its Use and its Impact [11].

The model that is compared as generic has been called the 5-Vowel Model (Fig. 1). This model emphasizes the use and impact of IS, as well as the user satisfaction and net benefits, considering that the quality of the system is directly related to these variables [38], Thus, all Information Technology (IT) artifacts, in most cases represented by software projects, usually evaluated by specific metrics, require a perspective from the different sciences.

The new European regulations will require companies to carry out an impact assessment by setting a series of criteria for carrying it out [12].

Therefore, all IT artifacts, in most cases, represented by software projects [13], which are generally evaluated by specific metrics, require a perspective from the different sciences, which leads to think that IS need an interdisciplinary vision [14].

The inclusion of the individual and organizational Objective and concepts such as effectiveness based on effectiveness (fulfillment of Objectives) and Efficiency

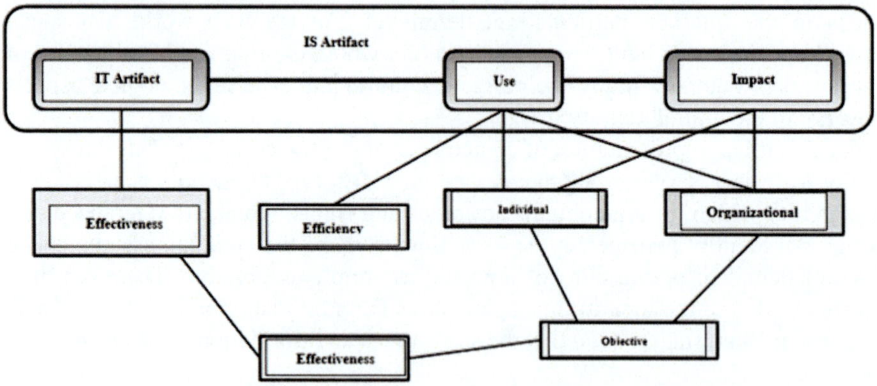

Fig. 1 General model

(input/output ratio) aims to make a representation of the Information systems from the perspective of their quantitative and qualitative analysis.

The following criteria proposed in other reports and studies of the Spanish Data Protection Agency are adopted, WG29:26: (1) The relationship between the purpose for which the data were taken and the purpose of the current Model, (2) The context in which the data were obtained and the expectations of the subjects, (3) The nature of the data and its possible impact, (4) The security measures that are applied such as the imposition of data use conditions when the data are transferred to third parties. In addition, an explicit recognition of the concept of Privacy by Design and Privacy by Default is made. The solution proposed is also the application of an "ethical privacy" model [30].

This paper aims to provide a model (Fig. 2) for both public and private organizations to manage the privacy security of sensitive data, in order to strengthen data protection according to current standards.

A new business model is proposed based on the empowerment of individuals. The World Economic Forum proposes two phases to develop the new model. The first phase is based on this creation of context-differentiated permissions, the second phase consists of information management, which implies a paradigm shift in data management with respect to current models.

In this way, the focus is on the use of the data and not on the collection, and the intention is to give a more relevant role (and greater responsibilities) to the companies and organizations that will make use of the data. The fundamental advantage will be to allow the measurement of effectiveness and efficiency, taking into account the effectiveness and how it is used and impacts on privacy [14].

Privacy poses risks that must be well managed professionally in a way similar to the risk category. Organizations handling personal data must monitor their ongoing

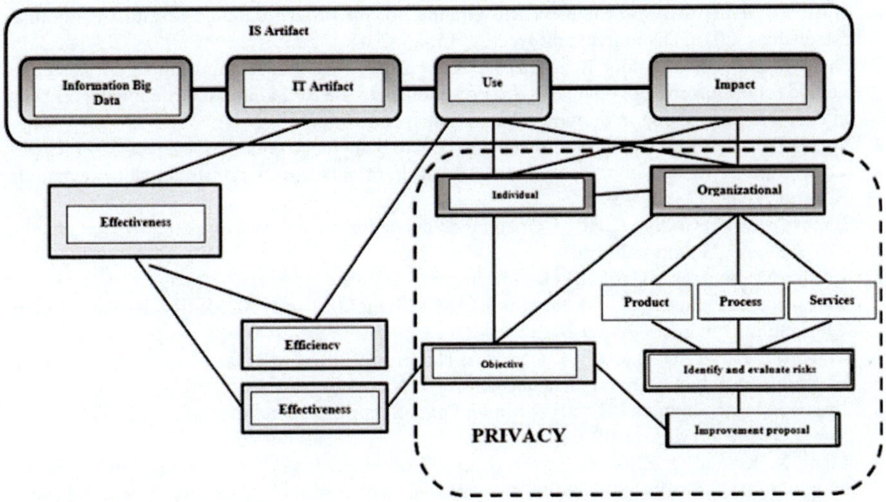

Fig. 2 Model of privacy

processes, as they are dealing with data of customers, employees or the general public and such processing may affect fundamental rights [15].

6 Conclusion

This paper combines experience from different areas of the project to provide valuable information on what the large data are doing, what they can do and what they should be allowed to do [16]. Increased regulation and legislation on privacy is also impacting IT environments [17]. The adoption of models and standards facilitates the rapid implementation of good procedures and helps avoid unnecessary delays in the development of new approaches [18]. All companies would need to adapt to the use of models and set standards to suit their individual requirements [39]. Each organization should establish its own management structure and reflect in all of them the recommendations that are most useful. The use of standards helps to comply with laws, regulations, contractual agreements and policies and to gain competitive advantages over other organizations [15]. The generic model presented for Big Data privacy management will allow the definition of more specific models according to the context in which the data are used.

References

1. Joyanes L (2014) Big Data: análisis de grandes volúmenes de datos en organizaciones. Barcelona: Marcombo. Ediciones Técnicas 2014. http://www.marcombo.com/Big-data_isbn9788426720818.html
2. Brasil. IBGE I Brasil em Síntese I Rio Grande do Sul I São Lourenço do Sul I Panorama. Estadísticas. 2016. Disponível em: Access: 15 jan. 2018
3. Chen Q, Zobel J, Verspoor K (2015) Evaluation of a machine learning duplicate detection method for bioinformatics databases. In: Proceedings of the ACM ninth international workshop on data and text mining in biomedical informatics. ACM, pp 4–12
4. Apraéz BE (2015) La responsabilidad por producto defectuoso en la Ley 1480 de 2011. Explicación a partir de una obligación de seguridad de origen legal y constitucional. Revista de Derecho Privado 28:367–399
5. Bueres A (2014) Código Civil y Comercial de la Nación analizado, comparado y concordado, 1ra Ed Bs. As. Hammurabi (Ed)
6. Navarro M, Becerra MC (2013) Gestión Integral de Infraestructuras Criticas en las Organizaciones Locales alineados a las Normas IRAM ISSO 27.001 y 27002. WSI—II Workshop de seguridad informática CACIQ 2013
7. Triguero I, Peralta D, Bacardit J, García S, Herrera F (2015) MRPR: a MapReduce solution for prototype reduction in big data classification. Neurocomputing 150:331–345
8. Burgos Salazar, Campos PG (2017) Modelo para Seguridad de la Información en TIC. http://ceur-ws.org/Vol-488/paper13.pdf
9. Zhang X, Khanal U, Zhao X, Ficklin S (2017) Making sense of performance in inmemory computing frameworks for scientific data analysis: a case study of the spark system. J Parallel Distrib, Comput

10. Borghello C, Temperini MG (2012) 41 JAIIO. Suplantación de Identidad Digital como delito informático en Argentina. www.41jaiio.sadio.org.ar/sites/default/files/7_SID_2012.pdf
11. La gestión de la identidad digital: Una nueva habilidad informacional y digital. BID. Universidad de Barcelona. http://bid.ub.edu/24/giones2.htm 2010
12. Bengio Y, Courville A, Vincent P (2013) Representation learning: A review and new perspectives. IEEE Trans Pattern Anal Mach Intell 35(8):1798–1828
13. Silva J, Cubillos J, Villa JV, Romero L, Solano D (2019) Preservation of confidential information privacy and association rule hiding for data mining: a bibliometric review. Procedia Comput Sci 151:1219–1224; Talavera-Llames R, Pérez-Chacón R, Martínez-Ballesteros M, Troncoso A, Martínez-Álvarez F (2016) A nearest neighbours-based algorithm for big time series data forecasting. In: Proceedings of the international conference on hybrid artificial intelligence systems, 2016, pp 174–185
14. Márquez Buitrago F (2016) Aplicación de la ley estatutaria 1581 de 2012 a la red social facebook en colombia. Revista De Derecho Comunicaciones Y Nuevas Tecnologías (15):1–31. 10.15425/redecom.15.2016.04
15. Política de Big Data (2018) Revista Dinero. https://www.dinero.com/pais/articulo/colombia-ya-tiene-politica-publica-debig-data/257479
16. DNP (2018) Documento CONPES 3920, Politica Nacional de Explotación de Datos (Big Data). https://colaboracion.dnp.gov.co/CDT/Conpes/Econ%C3%B3micos/3920.pdf
17. Puldain Salvador V (2017) El futuro marco legal para la protección del acceso a los datos. Revista Ibero-Latinoamericana de Seguros 26(47); Supriyadi D (2017) Personal and Non-Personal Data, In the context of Big Data. Tesis, Tilburg. 123; Sriramoju SB (2017) Introduction to Big Data: infrastructure and networking considerations. http://one.com.vn/sites/default/files/fileattached/catalog/introduction_to_big_data__infrastructure_and_networking_considerations.pdf
18. Daros WR (2017) ¿ Qué es un marco teórico?. Enfoques 14(1 y 2):73–112
19. Daries JP, Reich J, Waldo J, Young EM, Whittinghill J, Ho AD, Chuang I (2014) Privacy, anonymity, and big data in the social sciences. Commun ACM 57(9):56–63. https://doi.org/10.1145/2643132
20. Garriga Domínguez A (2016) Nuevos retos para la protección de datos personales. En la Era del Big Data y de la computación ubicua, Dykinson
21. Zhang X, Leckie C, Dou W, Chen J, Kotagiri R, Salcic Z (2016) Scalable local-recoding anonymization using locality sensitive hashing for big data privacy preservation. In: Proceedings of the 25th ACM international on conference on information and knowledge management. ACM, pp 1793–1802. https://doi.org/10.1145/2983323.2983841
22. Engan M (2017) Big Data and GDPR. Tesis de maestría, University of Stavanger. https://brage.bibsys.no/xmlui/handle/11250/2467489
23. Forgó N, Hänold S, Schütze B (2017) The principle of purpose limitation and Big Data. In: New technology, Big Data and the law. Springer, Singapore, pp 17–42
24. Gayo MR (2017) Big data: hacia la protección de datos personales basada en una transparencia y responsabilidad aumentadas. Revista De Derecho Comunicaciones Y Nuevas Tecnologías (17):1–24. 10.15425/redecom.17.2017.09. 59
25. Gellert R (2018) Understanding the notion of risk in the General Data Protection Regulation. Comput Law Secur Rev 34(2):279–288. Dhttps://doi.org/10.1016/j.clsr.2017.12.003 60
26. Viloria A, Viviana Robayo P (2016) Virtual network level of application composed IP networks connected with systems—(NETS Peer-to- Peer). Indian J Sci Technol 9(46). https://doi.org/10.17485/ijst/2016/v9i46/107376
27. Moreno A, Redondo T (2016) Text analytics: the convergence of Big Data and Artificial Intelligence. Int J Interact Multimed Artif Intell 2016
28. Benantar M (2006) Access control systems-security-identity management and trust models. Springer, New York
29. Féliz-Sánchez A, Calvo-Manzano JA (2015) Comparison of models and standards for implementing IT service capacity management. www.redalyc.org/pdf/430/43038629008.pdf

30. Frankel RM, Joos PR, Weber JP (2002) An empirical investigation of pre-earnings announcement quit period. Financial Accounting eJournal. https://papers.ssrn.com/sol3/papers.cfm?abstract_id=304486
31. Gaitán-Angulo M, Cubillos Díaz J, Viloria A, Lis-Gutiérrez JP, Rodríguez-Garnica PA (2018) Bibliometric analysis of social innovation and complexity (Databases Scopus and Dialnet 2007–2017). In: Tan Y, Shi Y, Tang Q (eds) Data mining and Big Data. DMBD 2018. Lecture notes in computer science, vol 10943. Springer, Cham
32. Silva J, Solano D, Fernandez C, Romero L, Villa JV (2019) Privacy preserving, protection of personal data, and Big Data: a review of the Colombia case. Procedia Comput Sci 151:1213–1218
33. Jacobson I, BoochG, Rumbaugh J (2000) El proceso unificado del desarrollo de software. Pearson
34. Information Systems/ spring 2003, vol 19, no 4, pp 9– 30. © 2003 M.E. Sharpe, Inc. 0742–1222/ 2003
35. Arce I (2014) Seguridad Tic. Desafíos y oportunidades para emprendimientos de base tecnológica en Argentina. Fundación Dr. Manuel Sadosky 2014
36. Estévez GP-T (2015) Estudio sistemático de literatura de metodologías para la obtención de requisitos de privacidad. http://www.dit.upm.es/posgrado/doc/TFM/TFMs2014-2015/TFM_Gonzalo_Perez-Tome_Estevez_2015.pdf
37. Gil González E (2016) Big Data-Privacidad y Protección de datos. Madrid 2016
38. Yan XF, Zhang DX (2013) Big data research. Comput Technol Develop 23(4):168–172
39. Bayern SJ (2009) Rational Ignorance, rational closed-mindedness, and modern economic formalism in contract law. Calif Law Rev 97(3):943–973. http://dx.doi.org/https://doi.org/10.15779/Z38TD7R
40. Hox JJ, Boeije HR (2005) Data collection, primary versus secondary. Encyclopedia of social measurement, 1, 593. https://dspace.library.uu.nl/bitstream/handle/1874/23634/hox_05_data+collection,primary+versus+secondary.pdf?sequence=1

The Goals Programming as a Tool for Measuring Sustainability of Agricultural Production Chains of Rice

Ana María Echeverria, Jesús García Guliany, and Ernesto Steffens Sanabria

Abstract Proposing coordination strategies for logistics networks implies, in addition to the formulation of operational proposals for the processes of supply, storage and distribution, and the design of the network structure through which goods/services, money and information will flow; the analysis of the conditions of the economic and productive context in which the logistics activity is framed, and the articulation of these to the proposals made, to make the productive chain more efficient and meet the requirements of the client, under the restrictions imposed by the environment. This paper analyses the factors that determine the production, postharvest handling and negotiation processes, as key elements that influence the operation of the rice logistics networks in Vatapá (Brazil), and that are conditioning factors for the design of coordination strategies. The result of the analysis allows for the identification of actions that decision-makers must take to impact the efficiency of logistics networks.

Keywords Coordination · Logistics · Analysis of main components · Agricultural supply networks · Rice

1 Introduction

Public policy decision-makers in developing countries such as Brazil have admitted the lack of willingness on the part of governments to support the development and

---

A. M. Echeverria (✉)
Universidad de La Costa, Barranquilla, Colombia
e-mail: aecheverria@cuc.edu.co

J. G. Guliany
Universidad Simón Bolívar, Barranquilla, Colombia

E. S. Sanabria
Corporación Universitaria Latinoamericana, Barranquilla, Colombia
e-mail: ssanabria@ul.edu.co

R. Patgiri et al. (eds.), *Proceedings of International Conference on Big Data, Machine Learning and Applications*, Lecture Notes in Networks and Systems 180,
https://doi.org/10.1007/978-981-33-4788-5_13

use of new technologies and knowledge to strengthen the marketing of food products in the country and abroad [1]. This lag in state control is due in part to the low levels of organization, integration and coordination of some of the country's production chains, especially those that are territorially isolated from central government institutions. In addition to infrastructure limitations, these chains have to overcome factors of violence, low levels of education and little knowledge of the markets, causing the production and commercialization of commodities to be carried out in a disorderly way and without following any normative guidelines, with the objective of sustaining the day-to-day life and not increasing the levels of competitiveness, considered as meeting the requirements of differentiation, quality, health and safety [2, 3].

Of the departments with the lowest levels of competitiveness, as classified by ECLAC, Vatapá is the one with the highest participation in the country's agricultural and livestock production, with products such as rice (ranked 2 nationally), cocoa (ranked 2), mechanized dry rice (ranked 5), cassava (ranked 6), traditional corn (ranked 4) and live cattle (ranked 10) [4]. Despite its productive participation, the department's most serious weaknesses, which justify its position in the ranking, are framed by the lack of infrastructure for development (roads, technology, energy and logistical services), the scarcity of qualified human capital for the consolidation of productive agricultural and industrial activities, the economic dependence on oil exploitation, the problems of insecurity generated by the armed conflict and, above all, the weak development of productive chains [5].

The paradox of the department of Vatapá having a low level of competitiveness and, at the same time, high percentages of agricultural production participation, makes it urgent to have measures and strategies that make the source of development in the region emerge from comparative advantages to competitive advantages, based on food control that guarantees the quality, health and safety of the products offered both nationally and internationally [6].

Underlying this scenario, the following research questions arise: What are the characteristics of the current logistic and productive operations in Vatapá's rice production chains? What are the critical processes in the mentioned chains that have the greatest impact on the efficiency of the logistic system? How can we influence the level of competitiveness of the department of Vatapá through the rice production chains, allowing its efficient control in terms of quality, health and safety, and at the same time generating competitive advantages that open the doors to national and international markets?

The objective of this paper is to develop a framework for analyzing the factors that influence the logistical performance of rice chains in the production, postharvest management and negotiation processes. The study is based on the statistical technique of analysis of main factors and programming by goals, to identify, within the information of more than 200 surveys, the key elements to influence the proposal of coordination strategies for the improvement of efficiency levels.

2 Method

2.1 Sample Design

The project covers 37 villages in the municipalities of Vatapá and Moqueta in northeastern Brazil, with approximately 1,450 families engaged in agricultural production, which corresponds to 67% of the total population in the areas affected by the oil wells [7].

A total of 214 surveys were conducted among producers and intermediaries, who responded to 1200 requests for data, thus consolidating a matrix of 124,325 attributes that was later analyzed using Microsoft Excel and SAS® statistical software.

The determination of the sample corresponded to a nonprobabilistic or directed process, which considered, among other factors, the circumstances of public order, the traditional vocation of the land, the proportional inclusion of the paths of the study area and the important presence of rice crops [8].

The description of the findings was classified into three major aspects that frame the logistical operations of agricultural supply chains such as: production, postharvest and negotiation processes [9].

Production processes were taken into account as a source of logistics operations, analyzing crop quality, maintenance practices, sowing and harvesting [10]. In postharvest processes, activities carried out once the fruit is extracted from the plant were characterized, such as transportation, handling, storage, processing and cleaning. After the findings from the field work, the negotiation processes were considered to be a fundamental motor for pulling the chain toward the levels of competitiveness it is in [11].

2.2 Analysis of the Principal Components

Due to the 124,325 attributes of the information base, it was necessary to implement a statistical method that would simplify the analysis with the least possible loss of information and by studying the structural factors of each of the characterized processes. With this intention, the potential of the principal component analysis was employed.

The multivariate statistical method—principal component analysis (PCA)—is explained in [12] as a data reduction technique used to identify a small set of variables that represent a large portion of the total variance of the original variables. This technique is also used to identify latent dimensions in the data.

PCA computes linear combinations of the variables. The first combination of variables represents the largest variation in the sample; the second is the next largest variation independent of the first one and so on [13].

To study the relationships between correlated variables (which measure common information), the original set of variables can be transformed into another set of new interrelated variables (which do not have repetition or redundancy in the information) called the set of principal components [14].

Principal component analysis does not require the assumption of multivariate normality of the data, although if the latter is true, a deeper interpretation of the principal components can be given [15].

3 Results

The economic activity of rice cultivation in the department of Vatapá is mostly developed by small producers, who use two alternatives for planting crops: the first one has to do with the planting of cocoa and rice in association, where rice plays a very important role since it provides the transitional shade to the cocoa in the first three years of establishment. It also provides the producer with economic income that contributes to the sustainability of the cocoa crop while it begins its productive stage, and in the second case, rice is established as a monoculture.

The analysis of the results corresponded to the processes of production, postharvest management and negotiation of the rice supply chains. In the production processes, within the scope of the original information collection instruments, the elements described in Table 1 were investigated [16, 17].

Table 2 presents the components analyzed for the postharvest handling process [3, 19].

Table 3 presents the components analyzed for the negotiation process [7, 20].

Table 1 Components for the analysis of the production process

Components for the analysis of production activities
Experience in the productive activity
Number of families depending on the activity
Average farmer's salary
Number of plants with active harvest
Cost of starting cultivation
Time to obtain the first harvest Monthly expenditure to maintain one-hectare
Type of product qualities

Table 2 Components for the analysis of the postharvest process

Activity analysis components postharvest handling
Reasons for loss in sales
Form of packaging of the product for sale
Postharvest activities carried out
Modes and means of transportation
Place of product delivery Transport distances and times Storage modes and means
Waste and spoiled product logistics

Table 3 Components for the analysis of the negotiation process

Analysis components of the negotiation activities
Relationship with suppliers:
– Geographical location of suppliers
– Number of suppliers
– Forms of payment for inputs
Customer relations:
– Number of buyers of the product
– Product payment methods
– Ways and means of communication
– Reasons for selecting buyers

The simplification of these variables is possible through the technique of principal component analysis, for which the following factors are defined in the case of the production process [21, 22].

1. Number of plants
2. Area to be harvested
3. Initial cost
4. Time it takes to obtain the first production
5. Monthly expenditure to maintain one hectare

3.1 Production

The study is based on the analysis of the dispersion of the evaluated factors, which, as shown in Fig. 1, has an acceptable level of variability of the data with a tendency to centrality, except in the case of the quantities of plants on the premises, which warns of the heterogeneity in terms of the production size of the analyzed sample.

Of the identified factors, according to the criterion of variance, those that explain a sufficient percentage of the variance are selected, which can be between 80 and 90%.

Under this criterion, factors 1, 2 and 3 are selected that explain about 80% (79.98%) of the variance of the data analyzed, as shown in Table 4.

The selection of the factors allows the identification of the axes for the biplot in which the correlations between the components are verified.

In the case of the biplot graph with axes 1 and 3 in Fig. 2, the close link between the initial investment cost for rice cultivation (Dim 3), the time it takes to obtain the first production (V4) and the monthly expenditure to maintain the crop (V5) is evident. This correlation supports public policy strategies that encourage credit for farmers in the early stages of their crops in order to have a positive impact, from the perspective of the supply chain, on reducing monthly expenditure and waiting times for the first production.

Due to the dimension of the vector V2, it is not possible to conclude the correlation ratio with the analyzed factors. However, when reviewing the biplot graph with axes 2 and 3, a strong relationship is found between it (area to be harvested) and the initial

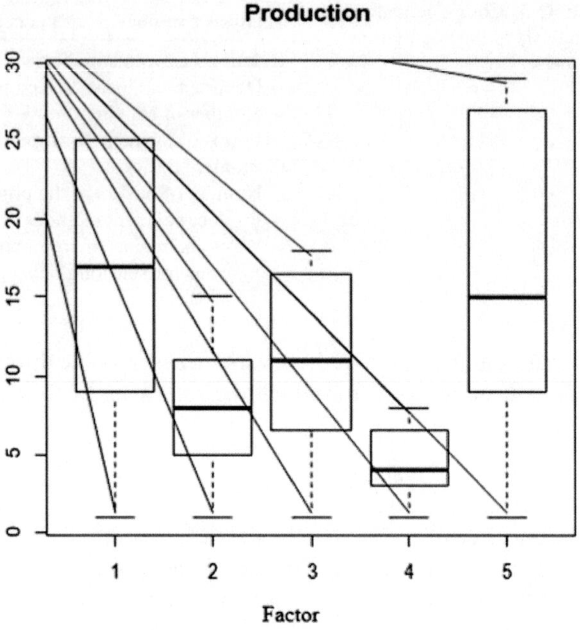

Fig. 1 Dispersion of the factors evaluated in the production process. Source: Own elaboration

Table 4 Accumulated inertia of the factors

Factors	Accumulated inertia of the factors		
	Eigenvalue	Percentage of variance	Accumulated % of variance
Factor 1	2.1041327	44.332547	42.08265
Factor 2	1.1152810	23.331458	64.38827
Factor 3	0.7411253	15.47123	79.9811
Factor 4	0.5714751	11.429502	90.64028
Factor 5	0.4679860	9.359721	100

Table 5 Accumulated Inertia of Factors

	Accumulated inertia of factors		
	Eigenvalue	Percentage of variance	Accumulated percentage of variance
Factor 1	1.4091827	47.58162	46.97276
Factor 2	0.9897890	33.58124	79.96572
Factor 3	0.6010283	20.03428	100

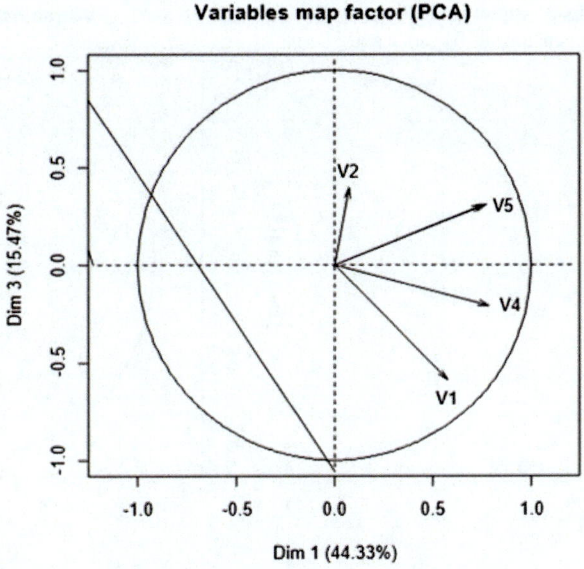

Fig. 2 Biplot in two dimensions of factors 1 and 3

cost to undertake the crop, allowing to extend the previous recommendation to the scope of the factor of area to be harvested.

3.2 Negotiation and Postharvest Management

Negotiation processes refer to the strategies used by the producer to purchase products within the production unit and from production to the external market. See Fig. 3.

The factors taken into account for the analysis of the negotiation process are.

1. Number of buyers
2. Distance between production site and delivery
3. Time it takes from the time it leaves the farm to the delivery point

The dispersion profile of the data obtained for each of the factors is presented in Fig. 4. The graph shows the heterogeneity of the characteristics of both the number of buyers and the time between production and delivery sites, elements that must be taken into account in the configuration of the network and the coordination strategies in biagent or multiagent relationships.

The selection of factors for the bidimensional analysis is done under the criterion of variance in which factors 1 and 2 represent about 80% of the behavior of the same.

An expected result of the correlation analysis of these factors is presented with the proximity evidenced in the behavior of the distance factors from the center of production to the point of delivery (Factor 2) and the time it takes to travel that distance (Factor 3), as presented in Fig. 5.

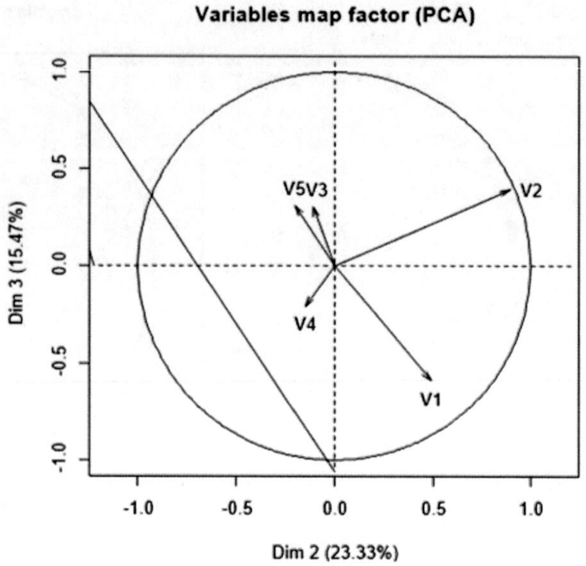

Fig. 3 Biplot in two dimensions of factors 2 and 3

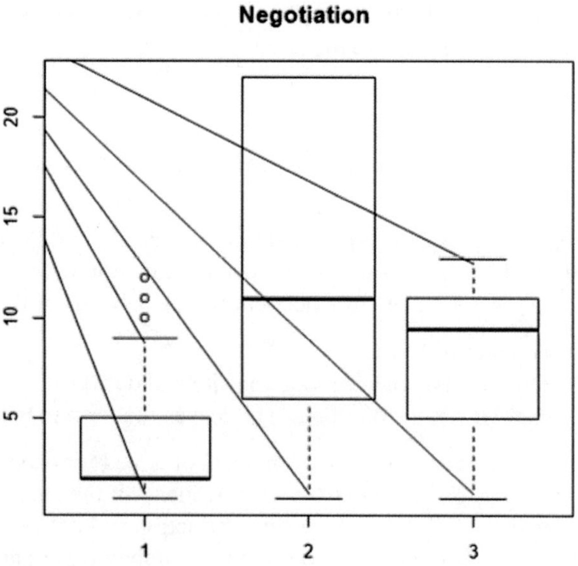

Fig. 4 Dispersion of factors assessed in the negotiation process

For a deeper understanding of this finding, the qualitative responses were analyzed. In terms of clients, most of the surveyed producers have commercial relationships with only one buyer (more often). When asked about this behavior, the producers expressed that the marketer who serves them is particularly characterized by the broad experience in rural agribusiness, the knowledge of the region, the degree of friendship they maintain over time and the economic capacity that marketers

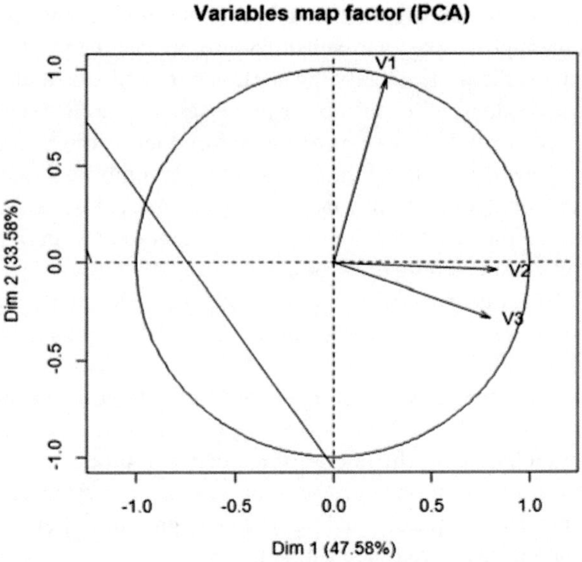

Fig. 6 Biplot with axes in factors 2 and 3

have to grant advance credit on the harvest to the producers without demanding any formality. In this way, the production obtained becomes the guarantee pledge. In addition, they consider that due to the low volumes produced in the production units, it is not necessary to maintain a commercial relationship with several traders.

Postharvest handling is also influenced by the place where the product is delivered to the intermediary, where 54.8% of producers deliver the product to their farm, 38.3% at a strategic point to the buyer and 2.55% carry out this activity at the municipal market or at the buyer's premises.

The rice marketer agrees in advance with the producers of a given area, the amount that each one estimates to sell, in order to establish the volumes to be mobilized. In this way, he defines the logistics related to the type of vehicle to be used, depending on whether the route is by land or river, to reach the production units. The grower, for its part, assumes the commitment to harvest the agreed-upon production, to move it from the lot to a visible place, in advance to avoid unnecessary delays in shipment, so that the buyer can bring forward the weighing of the bunches and the cancellation of the agreed-upon resources to the grower.

4 Conclusions

The analysis of main components for the case of production processes shows the importance of the initial cost of the crop as a factor strongly correlated with the components of quantity of plants, area to be harvested, time it takes to obtain the first production and monthly expenditure to maintain one hectare.

This result suggests an important effort in the design of strategies aimed at reducing this cost and subsidizing it so that a high level of investment allows the achievement of capacity levels (number of plants, area to be harvested, time for the first production) expected for the design of efficient networks.

Within the limitations of network models is the assumption of capacity levels that justify the operation of the same, these assumptions, according to the behavior of the main components, may lack factual accuracy if in the context there are no incentives and credit structures that allow producers to have high levels of production and low costs of sustaining the crop.

In addition to public policy strategies to strengthen agricultural credit, technical assistance should be considered to make efficient use of resources in the development of new crops.

With regard to the components of the negotiation and postharvest management processes, the design of strategies must consider, for the context under analysis, the characteristic of the relationships between producers and buyers, in which distance and transportation times are not a determining factor, as suggested by the classic models of allocation and network design, but rather the negotiation conditions and guarantees offered by a single buyer.

However, when analyzing the commercial dynamics that are evident in the different producing zones of the department, the producer negotiates individually with the offer of low volumes of products, which has repercussions on the fragile capacity of negotiation, a situation that is channeled by the intermediary in its favor by imposing the price of the products, causing constant fluctuations in them, affecting the economy of the primary link of the productive chain.

Taking into account solutions for the negotiation processes must be a priority for the agents in the Rice chain, especially for the producers, because given the characteristics, it is the client that marks the operation of the chain, and if the producers manage to establish negotiations in which concepts such as consolidation of the load, economies of scale, programmed deliveries, constant demand, production and the establishment of prices coherent with the dynamics of the market in general are privileged, the development of the sector will be evident, represented in profitability and sustainability for the entire chain.

Under these elements, mathematical models for the design of networks and coordination strategies among actors must take into account the characteristics of each context, the cost structures of each link in the chain, and even the idiosyncrasies of the population.

References

1. Abreu GB, de Carvalho JLV, Rangel PHN (2019) Contenido de zinc de los genotipos del arroz de la colección nuclear de la Embrapa. In Embrapa Cocais-Artigo em anais de congresso (ALICE). In: REUNIÓN ANUAL DEL PROGRAMA COOPERATIVO CENTROAMERICANO PARA EL MEJORAMIENTO DE CULTIVOS Y ANIMALES, 64., 2019, Tela Atlántida. Memorias Técnicas y Presentaciones. Tegucigalpa: IICA: Secretaria de Agricultura e Ganaderia de Honduras
2. Ramírez JJC, Osorio H, Parra-Peña R, CEPAL Bogotá (2007) Escalafón de la competitividad de los departamentos en Colombia. Bogotá: Naciones Unidas, Oficina de la CEPAL
3. Rubiano-Rodríguez JA, Sánchez-Doria T, Tofiño-Rivera R (2019) Estudio poblacional de Steneotarsonemus spinki (Acari: Tarsonemidae) sobre tres variedades de arroz sembradas en diferentes épocas. Revista Colombiana De Entomología 45(1):e7809–e7809
4. Blackburn J, Scudder G (2009) Supply Chain strategies for perishable products: the case of fresh produce. Product Operat Manag 18:129–137
5. Rong A, Akkerman R, Grunow M (2011) An optimization approach for managing fresh food quality throughout the supply chain. Int J Prod Econ 131:421–429
6. Romano P (2003) Co-ordination and integration mechanisms to manage logistics processes across supply networks. J Purchas Supply Manag 9:119–134
7. Silva J, Varela N, Lezama OBP, Hernández-P H, Ventura JM, de la Hoz B, Coronel LP (2019, July) Multi-dimension tensor factorization collaborative filtering recommendation for academic profiles. In International Symposium on Neural Networks (pp. 200–209). Springer, Cham
8. Klose A, Drexl A (2005) Facility location models for distribution system design. Eur J Oper Res 162:4–29
9. Zhang G, Habenicht W, Ludwig Spieß WE (2003) Improving the structure of deep frozen and chilled food chain with tabu search procedure. J Food Eng 60:67–79
10. Pfohl H-C, Köhler H, Thomas D (2010) State of the art in supply chain risk management research: empirical and conceptual findings and a roadmap for the implementation in practice. Logist Res 2:33–44
11. Amelec V, Carmen V (2015) Validation of a model for productivity evaluation for microfinance institutions. Adv Sci Lett 21(5):1610–1614
12. Mula J, Peidro D, Díaz-Madroñero M, Vicens E (2010) Mathematical programming models for supply chain production and transport planning. Eur J Oper Res 204:377–390
13. Lucas MT, Chhajet D (2004) Applications of location analysis in agriculture: a survey. J Operat Res Soc 55:561–578
14. Melo MT, Nickel S, Saldanha-da-Gama F (2009) Facility location and supply chain management—a review. Eur J Oper Res 196:401–412
15. Valcárcel-Aguiar B, Murias P (2019) Evaluation and management of urban liveability: a goal programming based composite indicator. Soc Indic Res 142(2):689–712
16. Omrani H, Valipour M, Emrouznejad A (2019) Using weighted goal programming model for planning regional sustainable development to optimal workforce allocation: an application for provinces of Iran. Soc Indic Res 141(3):1007–1035
17. Kim WC, Kwon DG, Lee Y, Kim JH, Lin C (2019) Personalized goal-based investing via multi-stage stochastic goal programming. Quant Fin, 1–12
18. Lachhwani K (2019) On multi-level quadratic fractional programming problem with modified fuzzy goal programming approach. Int J Operat Res 37(1):135–156
19. Amelec V (2015) Validation of strategies to reduce exhausted shelf products in a pharmaceutical chain. Adv Sci Lett 21(5):1403–1405

20. Viloria A, Robayo PV (2016) Inventory reduction in the supply chain of finished products for multinational companies. Indian J Sci Technol 8(1)
21. Barbosa L (2007) Feijão com arroz e arroz com feijão: o Brasil no prato dos brasileiros. Horizontes Antropológicos 13(28):87–116
22. Costa SMAL, de Souza Ferreira Filho JB (2019) LIBERALIZAÇÃO COMERCIAL NO BRASIL E INTEGRAÇÃO NOS MERCADOS DE COMMODJTIES AGRÍCOLAS: OS MERCADOS DE ALGODÃO, MILHO E ARROZ. Revista de Economia e Sociologia Rural 38(2):41–70

Multidimension Tensor Factorization Collaborative Filtering Recommendation

Harold Neira, Jesús García Guliany, and Luis Cabás Vásquez

Abstract A recommended system (RS) seeks to predict the preference that a user would give to a product in use, provides personalized information for the identification of articles, generating suggestions that are beneficial and agile for the search of the required items or activities. The user can accept the recommendations by providing information that is stored in a database, and generates new suggestions. These systems are used in the most prominent platforms such as websites and social networks. These information filtering techniques focus on the main properties and characteristics of items and users. This paper presents an analysis of the recommended systems and the components involved in the development of their functions. It shows an individual approach to filtering techniques, classification of RSs, possible combinations of filtering techniques and finally the conclusions are obtained in the analysis of the Recommended Systems.

Keywords Recommendation · Filtering · Items · Multidimensional tensor · Content-based

1 Introduction

Today, the Internet is causing a fundamental transformation in society. Thanks to the Internet, there is free online access to all kinds of encyclopedias and dictionaries, and news from all over the world can be consulted. Hundreds of items can be purchased, and a number of services can be paid for. The possibilities offered by the Internet

H. Neira (✉)
Universidad de La Costa (CUC), Calle 58 # 55-66, Atlántico, Barranquilla, Colombia
e-mail: hneira@cuc.edu.co

J. G. Guliany
Universidad Simón Bolívar, Barranquilla, Colombia
e-mail: jesus.garcia@unisimonbolivar.edu.co

L. C. Vásquez
Corporación Universitaria Latinoamericana, Barranquilla, Colombia
e-mail: lcabas@ul.edu.co

R. Patgiri et al. (eds.), *Proceedings of International Conference on Big Data, Machine Learning and Applications*, Lecture Notes in Networks and Systems 180,
https://doi.org/10.1007/978-981-33-4788-5_14

are practically infinite and the greatest limitation is the time spent discovering what one needs or simply looking for something one [1].

As a result of this obstacle, people need the assistance of experts to point out, in a practical and quick way, their preferences and available possibilities. To help users deal with the large amount of information, many companies have implemented recommendation systems to guide their users. Research in the field of recommendation systems has been ongoing for almost half a century, but interest still remains high due to the breadth of applications [2].

Recommendation systems arise as a result of the excess of information exposed in various areas, which limits users to determining which products (commercial articles, television content, educational courses, etc.) are relevant to them. As its name indicates, this type of system is responsible for recommending or suggesting to users the products related to their preferences, among the many options available [3, 4].

A number of means have been proposed that are used as a basis for recommendation systems. This study shows an analysis of the filtering technique that uses methods to retrieve information, whose processes use different filtering algorithms [5] and their importance to produce recommendations provided to the user in relation to their tastes and preferences. The present paper is structured as follows: Sect. 2 presents definition and detail in particular of the filtering elements and techniques of RS; Sect. 3 contains a matrix with the classification of techniques used in the recommendation of products; a table presenting the possible combinations of filtering techniques; and finally Sect. 4 specifies the conclusions and recommendations obtained from the analysis.

2 State of the Art

It is necessary to know how to select an alternative that contributes with knowledge to the search and obtaining the required information. The Recommended Systems are an alternative to reach knowledge without having to search among all the existing possibilities. There are several techniques and software tools that are useful to make suggestions and recommendations to the user in the decision-making process.

For this purpose, an RS collects and analyzes users' preferences, usually on a website, stores, social networks, music or movie selection or broadcast sites, etc. The basic premise of an RS is that users with similar activities will continue to share preferences in the future. When a user is suggested services or tasks that other users with similar criteria have selected in advance, the level of success is likely to be higher than suggestions based on nonpersonalized preferences [6].

The main function of the recommended systems depends on the objectives that are set at the time of their definition and implementation; but in general terms, they tend to use three types of objects: items, terms that are to be suggested; users, types of users that order the requirements and priorities of a user within the system; and transactions, interrelations between the user and the system that are implicitly or explicitly collected [7, 8].

2.1 RS Based on Content

The study of content-based RS is supported by information retrieval, refining the methods of data collection. The systems focus on recommending articles that contain informative text, such as documents, websites (URLs), see [11, 12].

2.2 Collaborative Filtering Technique

Collaborative filtering is one of the most used techniques in recommendation systems. It is focused on the prediction and recommendation of elements based on the collection of preferences of many users. It is based on the premise that if two or more users have similar characteristics (behavior regarding their votes) they should have similar preferences. This measure of similarity is quantified to associate a group of collaborators to each user.

There are two trends for collaborative filtering algorithms: those based on memory and those based on models. Those based on memory make predictions from a database in which users, multimedia content and voting with their respective relationships are stored. With these data, the valuation history of the collaborators of the current user is obtained and a list of audiovisuals is inferred to be recommended. Those based on models create a model of the users and their valuations, from this the expected value for each multimedia is calculated based on the valuations made by the collaborators. It is based on learning algorithms [13].

The memory-based collaborative recommendation algorithms establish similarity relationships between users or between elements. These algorithms are based on mathematical formulas to quantify the similarity between two of these categories. The collaborative filtering technique presents several problems, among these is the scarcity of data due to the need for a large number of users performing similar content assessments, these data must provide a matrix of scores that allow neighborhood calculations, predictions and recommendations with a high reliability value. Another problem is the scalability of the system: given because the computational cost of the algorithms used to calculate the nearest neighbors grows as the corpus of data increases.

Based on the operation of this technique, multimedia content is recommended to the active user if one of their collaborators has made satisfactory ratings on it, when new content is published on the platform it is not recommended until it has received any rating, thus presenting the problem of the new element. When registering a new user on the platform, you do not have enough information about your preferences, because you have not made assessments of the published content, as a consequence you cannot be associated with any group and do not receive recommendations [14].

Collaborative filtering is a technique for making predictions about a user's preferences, based on information about the tastes of different users. The underlying assumption is that, if person A has the same opinion as person B on some issues, then A is more likely to have similar opinions as B on different issues. There are two popular forms of collaborative filtering: user-based and item-based [12], see [13, 14].

There are several techniques used in the recommendation systems, among the most addressed by the scientific community are:

- Content-based recommendation systems: it recommends audiovisual materials similar to those that the user liked, depending on their characteristics [7].
- Collaborative filtering recommendation system: recommends multimedia content that has had satisfactory votes or ratings and that has not been seen by the user, but by users who share similar preferences to this [8].
- Knowledge-based recommendation systems: make audiovisual suggestions based on inferences about users' needs and preferences [9]. These systems are governed by knowledge of the environment in which the recommendations are generated and are provided by specialists in the field.
- Usefulness-based recommendation systems: it uses a utility formula that directly intervenes in the recommendation process and that includes variables such as availability, quality of content and formats.
- Demographic recommendation systems: it makes recommendations based on geographic information, which allows the creation of geolocalized groups of users based on their audiovisual consumption.

Each of the previous techniques has its own strengths, but they present problems such as the high dependence on knowledge of the environment or the requirement of a large amount of data that characterize the materials they recommend, to name a few; to minimize these problems and increase the reliability of the recommendations, hybrid systems arise. These systems are formed by two or more recommendation techniques based on various hybridization criteria. Hybrid systems have high precision indicators in the recommendations, but they are complex in their development.

2.3 *The Proposed MD-TFCF Mechanism*

This section presents the framework of the Multi-Dimension Tensor Factorization Collaborative Filtering (MD-TFCF) approach (Fig. 1).

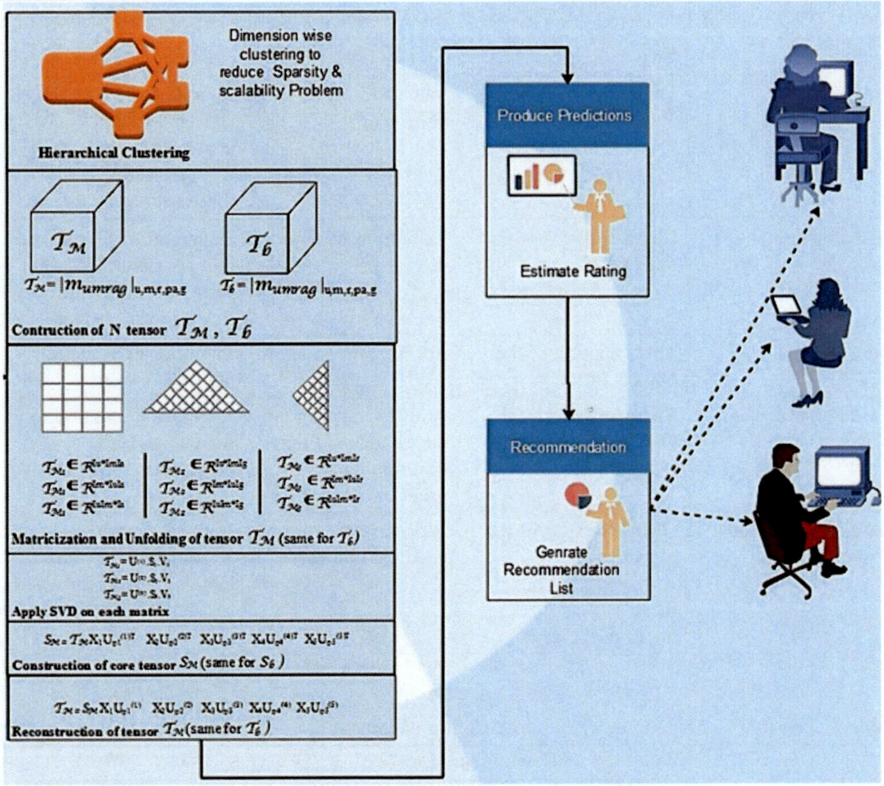

Fig. 1 (MD-TFCF) Framework, based on [15]

3 Analysis of Filtering Techniques

The RS that have been initially reviewed have different characteristics depending on the filtering techniques. Table 1 presents a summary of the techniques; assuming that I is the set of objects on which recommendations can be made; U is the set of users whose preferences are known; u is the user for whom recommendations are to be generated; and i is an object for which the preferences of u, $i \in I'$ with $I' \subseteq I$ [16].

This approach focuses on recommendations that combine information from different sources, as these are the most commonly implemented and have the greatest potential for solving the cold start conflict, seven types have been identified [17–20].

Table 2 shows the combination of the five recommendation approaches.

A = Combination of characteristics, B = Increase of characteristics.

C = Collaborative, D = Content-based, E = Demographic, F = Knowledge-based.

Where:

Redundant	
Not possible	
There is implementation	

Table 1 Recommendation techniques [16]

Techniques	Characteristics		
Collaborative	Valuations by U of the objects of I	Valuations by u of the objects of I	Identifies users in U similar to u and predicts the rating of i from the ratings of those users
Based on content	Characteristics of the objects of I	Valuations of objects of I	Generates a classifier that adjusts the behavior of the ratings of u and uses it in i
Demographic	The valuation of the objects of I	Information about u	Identifies users who are similar to u
Based on utility	Characteristics of the objects of I	A utility function on the objects in I that describe the preferences of u	It applies a function to the objects and determines the order of i
Based on knowledge	Characteristics of the objects of I. Knowledge of how these objects meet the needs of the user	A description of the needs or interests of u	It interferes with a relationship between i and the needs of u

Table 2 Possible combinations of recommendation systems [20]

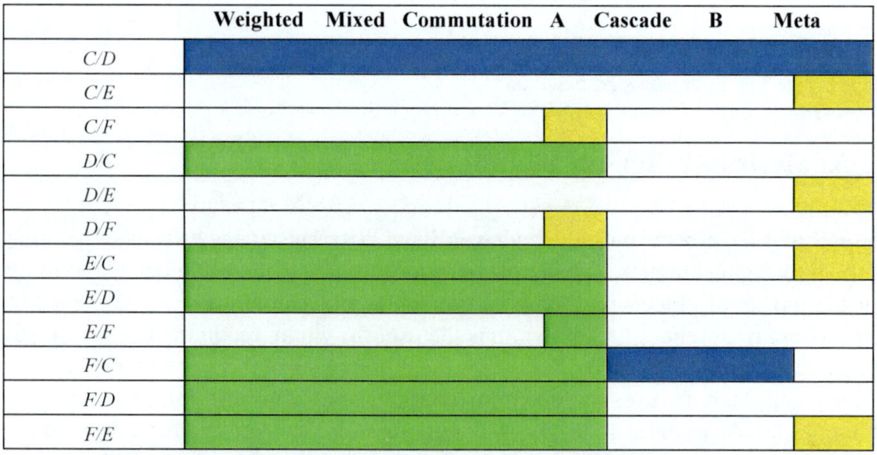

One way to build hybrid recommendation systems is to implement separate collaborative and content-based systems. Therefore, two different scenarios are possible. The outputs (ratings) obtained from recommendation systems can be combined into a final recommendation using either a linear combination of ratings [21]. Alternatively, we can use one of the recommendations, at any given time, to choose the best one, based on some "quality" metric [15, 22–24].

4 Conclusions

Recently, the recommended systems reached a significant rise in the field of recommendation, because multiple projects based on content and collaboration were combined to form the so-called hybrid methods. In this way, several systems were developed to offer services to users, such as e-commerce and websites. These systems used filtering techniques and search algorithms to generate suggestions to consumers. However, although the progress of RS is considerable, at present the RS reviewed in this paper still require improvements to make the recommendation methods more effective in an increasing scale of new systems.

This paper reviews the seven types of combinations that refer to RS techniques: weighted, switching, mixed, combination of characteristics, augmentation, cascade and meta-level. In Table 2 of Chapter 3, the 53 possible combinations between the recommended systems are determined, the feasibility between them is identified, i.e., whether the combination is possible, whether they are redundant or whether it has been implemented. There are limitations to the current recommendation techniques that could be improved by giving more importance to users and items, seeking support in the various criteria and gathering information to make way for a recommendation process that is more flexible than those that have existed so far.

References

1. Joyanes L (2014) Big data: análisis de grandes volúmenes de datos en organizaciones. Barcelona: Marcombo. Ediciones Técnicas. La gestión de la identidad digital: Una nueva habilidad informacional y digital. BID. Universidad de Barcelona https://bid.ub.edu/24/giones2.htm 2010
2. Silva J, Cubillos J, Villa JV, Romero L, Solano D (2019) Preservation of confidential information privacy and association rule hiding for data mining: a bibliometric review. Proc Comput Sci 151:1219–1224. R
3. Solis Jacqueline, Chacón-Rivas Mario Garita Cesar (2014) Agente Híbrido Recomendador de Objetos de Aprendizaje. Obtenido de Research Gate: https://www.researchgate.net/publication/270341748_Agente_Hibrido_ Recomendador_de_Objetos_de_Aprendizaje
4. Tovar L, Montoya J, Martelo R (2018) Sistema ecléctico de filtrado de información basado en inteligencia computacional para recomendación de atractivos turísticos del Caribe Colombiano. Recuperado el 11 de 07 de 2018, de https://revistas.sena.edu.co/index.php/LOG/article/download/1521/1692
5. Viloria A, Viviana Robayo P (2016) Virtual network level of application composed IP networks connected with systems - (NETS Peer-to- Peer). Indian J Sci Technol 9(46). doi:https://doi.org/10.17485/ijst/2016/v9i46/107376

6. Gonzaga-Perez L (2009) Modelos de recomendación con falta de información. Retrieved in 11 de 07 de 2018, de Catálogo del Centro de Documentación y Publicaciones del Servicio de Información y Documentación, Estudios y Publicaciones de la Secretaría General Técnica de la Consejería de Turismo, Comercio y Deporte de la Junta de Andalucía: https://www.juntadeandalucia.es/turismocomercioydeporte/publicaciones/36570.pdf
7. Isinkaye F, Folajimi Y, Ojokoh B (2015) Recommendation systems: principles, methods and evaluation. Egyptian Inf J, pp 261–273. Retrieved from: https://doi.org/https://doi.org/10.1016/J.EIJ.2015.06.005
8. Moreno A, Redondo T (2016) Text analytics: the convergence of big data and artificial intelligence. En international. J Int Mult Art Intel
9. Silva J, Solano D, Fernandez C, Romero L, Villa JV (2019) Privacy preserving, protection of personal data, and big data: a review of the Colombia case. Proc Comput Sci 151:1213–1218
10. Benitez R, Escudero G, Kanaan SM (2013) Inteligencia artificial avanzada. Barcelona: UOC
11. Bennet J, Lanning S (2007) The Netflix prize. Proceedings of KDD Cup and Worksho, pp 3–6
12. Linyuan L, Matúš Medo, Chi Ho Y, Yi-Cheng Z, Zi-Ke Z, Tao Z (2012) Recommender systems. Elsevier B.V, pp 1–49
13. López Puga JG (2007) Las redes bayesianas como herramientas de modelado en psicología. Anales de Psicología. Redalyc, pp 307–3016
14. Matich DJ (2001) Redes neuronales: conceptos básicos y aplicaciones. Cátedra de Informática Aplicada a la Ingeniería de Procesos-Orientación, Rosario
15. Lee J, Lee D, Lee YC, Hwang WS, Kim SW (2016) Improving the accuracy of top-n recommendation using a preference model. Inf Sci 348(1):290–304
16. Silva J, Varela N, Lezama OBP, Hernández-P H, Ventura JM, de la Hoz B, Coronel LP (2019, July) Multi-dimension tensor factorization collaborative filtering recommendation for academic profiles. In International Symposium on Neural Networks (pp. 200–209). Springer, Cham
17. Su X, Khoshgoftaar TM (2009) A survey of collaborative filtering techniques. Adv Artif Intell 4(1):1–12
18. Malik F, Baharudin B (2013) Analysis of distance metrics in content-based image retrieval using statistical quantized histogram texture features in the DCT domain. J King Saud Univ Comput Inf Sci 25(2):207–218
19. Isinkaye FO, Folajimi YO, Ojokoh BA (2015) Recommendation systems: principles, methods and evaluation. Egypt Inform J 16(3):261–273
20. Taneja A, Arora A (2018) Cross domain recommendation using multidimensional tensor factorization. Expert Syst Appl 92(1):304–316
21. Panniello U, Tuzhilin A, Gorgoglione M (2014) Comparing context-aware recommender systems in terms of accuracy and diversity. User Model User-Adap Inter 24(2):35–65
22. Zheng C, Haihong E, Song M, Song J (2016) CMPTF: contextual modeling probabilistic tensor factorization for recommender systems. Neurocomputing 205(1):141–151
23. Lops P, de Gemmis M, Semeraro G (2011) Content-based recommender systems: state of the art and trends. In: Ricci F, Rokach L, Shapira B, Kantor PB (eds) Recommender systems handbook, pp 73–105. Springer, Boston (2011). https://doi.org/https://doi.org/10.1007/978-0-387-85820-3_3
24. Gómez S, Zervas P, Sampson DG, Fabregat R (2014) Context-aware adaptive and personalized mobile learning delivery supported by UoLmP. J King Saud Univ Comput Inf Sci 26(1):47–61

A Recent Survey of SVD- and DWT-Based Digital Image Watermarking Theories and Techniques: A Review

Ranjeet Kumar Singh, Anil Kumar Singh, Naveen Kumar, and Ashwina Kumar

Abstract Digital image watermarking is one of the largest research fields, and it plays a vital role in digital information authentication and security. Based on the study of watermarking systems, the whole process may be segregated into the processes of watermark embedding and extraction. In this paper, a review of the theoretical and experimental analysis and performance measurement of representative digital image watermarking systems in spatial and transform domains is presented. The significant characteristics of the digital image watermarking scheme are robustness, capacity, imperceptibility, the security of the hiding place, and false-positive of the watermarking algorithm. Comparison of the different watermarking techniques such as the least significant bit (LSB), singular value decomposition (SVD), the discrete wavelet transform (DWT), and the discrete cosine transform (DCT) is presented in a tabular form. A detailed discussion along with an algorithm for digital image encryption/decryption through embedding and extraction is given. A brief comparison of their relative advantages and disadvantages is also noted in this paper.

Keywords Information hiding · Watermarking · Security · Attacks · Robust

---

R. K. Singh (✉) · N. Kumar · A. Kumar
DIT University, Dehradun, Uttarakhand 248009, India
e-mail: 2014rsca002@gmail.com

N. Kumar
e-mail: Naveen.it23@gmail.com

A. Kumar
e-mail: kumarashwina31@gmail.com

A. K. Singh
Assam Kaziranga University, Jorhat, Assam 785006, India
e-mail: singhanilkumar@zoho.com

R. Patgiri et al. (eds.), *Proceedings of International Conference on Big Data, Machine Learning and Applications*, Lecture Notes in Networks and Systems 180,
https://doi.org/10.1007/978-981-33-4788-5_15

1 Introduction

1.1 SVD Watermarking

Any distinct image may be expressed as an array of positive scalar entries that can be treated as a matrix. SVD is a linear algebraic transform for matrices that crumble the matrix into three submatrices:

$$M = USV^T = \begin{bmatrix} u_{1,1}, & K, & u_{1,n} \\ u_{2,1}, & K, & u_{2,n} \\ & M & \\ u_{n,1}, & L, & u_{n,n} \end{bmatrix} \begin{bmatrix} \sigma_{1,1}, & 0, & K, & 0 \\ 0, & \sigma_{2,2}, & K, & 0 \\ & M & O & M \\ 0, & 0, & L, & \sigma_{n,n} \end{bmatrix} \begin{bmatrix} v_{1,1}, & K, & v_{1,n} \\ v_{2,1}, & K, & v_{2,n} \\ & M & \\ v_{n,1}, & L, & v_{n,n} \end{bmatrix}^T$$

here, U and V are the unitary matrices, and S is a diagonal matrix having nonnegative values with $\sigma_{1,1} \geq \sigma_{2,2} \geq L \geq \sigma_{r,r} = \sigma_{r+1,r+1} = L = \sigma_{n,n} = 0$.

SVD has two prime features. First, when a small alteration is done to the image, there is very little change in the singular values of the image. The other more highlighted feature of SVD is the degree of freedom (DOF). The mapping of the cover image is done as follows:

$$DOF\ (S) \Rightarrow N$$
$$DOF\ (U) \Rightarrow N\,(N-1)/2, \ldots\ DOF(V) \Rightarrow N(N-1)/2 \ldots$$

There are various image watermarking algorithms available based on the singular value decomposition approach. SVD decomposed an image into *U*, *S*, and *VT* components. *S* is the diagonal element, and *U* and *V* being the orthogonal matrices. *S* component is used for embedding the watermark information because the *S* component contains only diagonal information, and compared with *U* and *V* components, this part's computational cost is low. Different SVD-based watermarking schemes are discussed in the paragraphs to follow.

Li et al. [1] came up with an algorithm that was developed on chaotic encryption and SVD. The authors segregated the cover image into 8 × 8 size blocks. They then enforced the DCT on each block and found out DCT coefficient matrices. Finally, these matrices were combined, and a new cover image was obtained. After completing this process, they applied SVD on the new cover image. The image was disintegrated into three components, viz. singular value matrix *S*, *U*, and *V* matrix. The *S* component was selected for inserting a watermark image. In this paper, chaos sequence was used for encryption of the watermark image.

In [2], a semibreakable digital image watermarking technique on SVD was discussed. The cover image was split into 8 × 8 size blocks, and by conducting SVD on every block, *S*, *U*, and *V* components were obtained. By using the quantization process, the authors inserted the binary watermark into the greatest singular value of the image. This approach could recover the watermark image without the use of the original image.

Agarwal et al. [3] explained an SVD-based invisible and nonblind digital image watermarking mechanism for colored images. In this scheme, identical or dissimilar watermark images could be inserted in the tri-color component, resulting in the rise of the robustness of the watermark. Here the authors at first converted the host image to its RGB color component. Then they applied the SVD on each color component and calculated the eigenvectors of each color component. These eigenvectors were added to its corresponding eigenvector of the RGB component of the watermark image.

In [4], an image watermarking technique that combined the application of the SVD and the DWT was discussed. The authors modified the conventional SVD and came up with what we know now as the shuffled SVD (SSVD). By applying DWT, the original image was divided into a vertical frequency band, a horizontal frequency band, a low-frequency band, and a high-frequency band (LL, HL, LH, and HH). The LL frequency band was again turned into N × N size blocks. The prime components of the watermarked image were implanted in several blocks of the LL frequency band with the help of SSVD.

Durga et al. [5] developed an algorithm on SVD on M × N blocks of both the cover image and the binary watermark image. By using SVD on the subblocks of the cover image as well as the watermark image, they found *U*, *S,* and *V* components. The *S* component of every nonoverlapped block of watermark image was inserted into the *S* component of corresponding nonoverlapped blocks of the cover image by using the Multiple Scaling Factors (MSFs) of the firefly system.

Farzad Golshan et al. [6] used a hybrid intelligence approach that relied on particle swarm optimization and genetic programming to make a tradeoff between imperceptibility and robustness. In this approach, the authors used intelligent methods in the DCT–DWT–SVD domain. Initially, every cover image was changed into the DCT domain, followed by a fragment of the DCT matrix selected and divided into four subbands with the help of a discrete wavelet transform. Finally, to insert the watermark, the *S* component of every subband was shaped perceptually by *S* element of the watermark image.

Hem Kandpal et al. [7] showed a hybrid image-based watermarking scheme using a combined approach wave atom transform and SVD. Here, the cover image was converted into five frequency subbands in the form of an array of cells by using the forward wave atom transform. They then applied SVD on the matrix that they obtained.

In [8], the scholars devised a unique watermarking technique for digital data, which was constructed on logistic and singular value decomposition. First, the original image was segregated into an 8 × 8 block, and then the grayscale watermark was scrambled using logistic mapping. After this, the authors applied SVD on each block to get the *U*, *S,* and *VT* and selected the singular value components for embedding the watermark image.

Zhang Nana [9] showed a blind digital image watermarking mechanism based on spatial images. First, the cover image was divided into M × M subblocks and applied SVD on every block. By scrambling the binary watermark image with the aid of a chaotic sequence, the watermark image was inserted into the singular value after completing quantification processing.

Soo-Chang Pei et al. [10] indicated a digital image watermarking approach in the spatial domain with the help of the SVD technique. This paper presented work on RGB color space in the spatial domain, and instantaneously inserted three binary watermark images into color cover images with small perceptual distortion. Moreover, the authentic user in the receiver required three secret keys to extract all the binary watermark images without any issue.

Suwen Zhu et al. [11] showed the robust image watermarking mechanism, which was based on the traditional W-SVD scheme. Here, the authors presented a novel, HVS-based image watermarking approach. Using the watermark strength regulation factor, the authors could further increase the robustness of this mechanism. Vital information was encrypted by using the Arnold transform and then revised into a watermark prototype. Then, the watermark data were inserted in the low-frequency component of the cover image.

The watermark synchronization problem is a big issue for digital security. Li Zhang et al. [12] proposed an algorithm for this problem. During the watermark insertion process, the watermark image was initially inserted into the cover image by using the SVD approach. Then, the Scale Invariant Feature Transform (SIFT) approach was employed to determine the feature-points of the watermark. In the watermark recovery process, the watermarked image, probably affected by geometric attacks, was matched with the original watermarked image to determine the uninformed three pairs of matched feature-points.

Radharani and Valarmathi [14] focused on a block-based SVD image watermarking approach. First, the watermark image was scrambled by using the Arnold transformation. Then, the original image was converted into its RGB components, and every color component was divided into N × N block sizes. Then they applied SVD on every block and select singular values for embedding watermark. The authors used Arnold transformation to provide multistage security.

SVD-based image watermarking is advantageous due to the minimum modification of pixel and reduced computational cost. SVD is used mainly to optimize the frequency domain approach to enhance their performance.

1.2 DWT Watermarking

DWT has a major role to play in image processing and watermarking. The transforms are built on several small waves known as wavelets. Moreover, the wavelet disintegrates the original signal into a wavelet transform coefficient, which carries the information about the position. Complete regenerated of the original signal is achieved by applying IDWT on these coefficients. DWT splits the original image into four segments: the lower resolution approximation component (LL), the vertical detail component (LH), the diagonal detail component (HH), and the horizontal component (HL). The lower resolution approximation component is a result of low pass filtering of the rows and columns, holding a vague description of the image. The diagonal detail component (HH subband) contains information regarding the diagonally high-frequency component. There have been precedence referring to the discussions on the use of LSB watermarking for image security and the role of noise in the process [15, 16]. In [17], the authors made use of DWT to employ a hybrid concept of random pixel exchange and compressive sensing to create an algorithm that can carry out compression and encryption at the same instance. In [18], the authors used level-2 DWT on the original image to find the LL2 subband and presented a block-based nonblind watermarking scheme that employed both DWT and SVD.

1.3 Arnold Cat Map

The chaotic map has a complex behavior that is used to provide multilevel security in the watermarking mechanism. Therefore, several watermarking schemes based on the chaotic map are available for providing security from unauthorized access. Normally, the 2D Arnolds Cat Map (ACM) finds application in image watermarking. In this technique, the position of pixels of a watermarked image is shuffled, resulting in the creation of a scrambled watermark image.

The mathematical function used to scramble the N × N square watermark image is explained in Eq. (1), where (X_n, Y_n) and (X_{n+1}, Y_{n+1}) are the pixel positions before and after ACM, respectively [19].

$$\begin{pmatrix} X_{n+1} \\ Y_{n+1} \end{pmatrix} = \begin{pmatrix} 1 & a \\ b & ab+1 \end{pmatrix} \begin{pmatrix} X_n \\ Y_n \end{pmatrix} \bmod N = A \begin{pmatrix} X_n \\ Y_n \end{pmatrix} \bmod N \quad (1)$$

Now, we discuss some watermarking techniques, which are based on discrete wavelet transform and Arnold cat map. Zhi-Feng Gu et al. [20] propounded a watermarking technique that was based on the distributing feature of the DWT coefficients. The authors presented an algorithm for image watermarking, which was developed on DWT and Chaos mapping. The watermark image was scrambled by logistic mapping, and the original image was converted by applying DWT. Low-frequency subbands were selected for embedding watermark image.

Zhang Guannan et al. [21] indicated a binary image watermark embedded into the original color image by using DWT. First, the original colored image was reformed into their RGB color components, and RGB color space was converted into the YUV color space post by dividing it into N × N sub-blocks. The watermark was inserted into the Y component. Manipulating the tremendous features of the wavelet transform, authors used the mean and variance of the subbands to modify the wavelet coefficients adaptively to insert the watermark image.

In recent years, Quick Response Code (QR Code) has found utility in image watermarking schemes. The QR code is a two-dimensional bar code that is used in the image processing scheme for enhanced the security of digital content. In [22], the hue of the original image was transformed from the RGB scheme to the YC_bC_r color scheme. The Y component was chosen, and the DWT transform was applied up to the third level. The authors chose the diagonal constituent part for the insertion of the watermarked image. They converted the text data into a QR code, and this QR code was used as a watermarked image.

Haiying Gao et al. [23] suggested improving data security through a DWT-based algorithm. Their paper presented an algorithm using DWT with alterable parameters to protect sensitive data. The parameters were used as a key to control the locations for secret data insertion.

Pramoun and Amornraksa [24] showed a new color image watermarking technique, which was based on coefficients modification in the LL component. First, the host image was transformed into its RGB color component, and then the blue component of the host color image was disintegrated into different bands by DWT. They then selected the LL subband to insert a watermark image bit with a fixed value to find the ideal watermark forte.

Vladimír Bánoci et al. [22] propounded a novel approach to insert secret messages in grayscale images through steganography. To ensure greater carriage in the confidential data and to preserve better visual features of the stego-image, insertion was implemented in the transform domain of DWT by changing the coefficients of transform in a suitable way.

Multilevel security is one of the most sought research fields in watermarking schemes. Zhou and Jin [25] presented a multiwatermarking scheme where three binary watermark images were inserted in a grayscale digital host image. To enhance the richness of the watermarked image and enhance the tenacity while recovery, the multiwatermark image was inserted. Three 2D watermark images were recreated into one 3D watermarking sequence. After that, the subimages of the main image were split into blocks, and those with good texture data got chosen. In the end, a multiwatermark image was inserted post changing the fractional values of the selected block pixels, which were based on the discrete operation law.

With the rapid development and applications of digital content over the internet, the protection of digital data against unauthorized users has become extremely challenging. The chaotic map is one of the schemes used to uplift the safety of the watermark information. Qiang et al. [26] wrote an article that was based on DWT and chaos theory. During the insertion of the watermark image, DWT was applied first, and then a low-frequency component was chosen as the inserting field, followed by the application of the chaotic sequence for the watermark encryption. The authors then applied DWT to the scrambled data and selected the low-frequency component. In the end, they inserted the low-frequency component in the cover image. It is a non-blind image watermarking technique. Here a unique image is required for recovery of the watermark image [27].

One logistic chaotic-based watermarking is explained by Sun et al. [28]. The authors made use of a chaotic logistic sequence and applied ACM to encrypt the watermark image. The safekeeping of the inserted watermarks was improved. Watermarks were inserted into the low-frequency module of the deep wavelet domain. We do not need the host image to recover the watermark image.

In [24], a mechanism of inserting an encrypted watermark image into the green element of the colored cover image is explained. The watermark insertion process is designed using a discrete wavelet transformation (DWT) domain. The RGB elements of the base image were broken down by using a level-2 DWT, i.e., LL, HL, LH HH, (k = 1, 2).

Liu and Chen [29] included the spatial domain approach and the transform domain approach in their study. The authors propounded a new hybrid scheme for the fusion of multifocus images. The error rate in the transform domain while selecting subband components was mitigated significantly. It also decreased the non-natural discontinuities formed in the spatial domain scheme. First, the wavelet transforms were applied to every input image. Then a focused region decision map was recognized. The fusion process is shown by this map, and the integrated components are regenerated to form the fused image.

Thongkor et al. [30] developed a watermarking technique to DWT coefficients modification. First, during watermark insertion, the blue components of the unique base image were decomposed by the DWT, and the LL subband along with the watermark image was inserted in this part. During the watermark recovery, the initial coefficient estimates in the LL component based on the main filter were used to recover the inserted watermark image.

Javier Molina-Garcia et al. [31] showed a digital watermarking approach comprising of two stages. The first one was used to protect the digital image, which was based on Daubechies DWT, halftoning and QIM approach. The next is applied for verification, detection, and recovery of tampered regions, through inverse DWT and inverse halftoning, median filtering.

Based on the above discussions, we now explain a few algorithm steps of DWT-based image watermarking.

Algorithm 1: Modifying the frequency component.

Embedding Process.

Step 1: Took the original picture.

Step 2: Applied DWT on it.

Step 3: Selected the watermark image.

Step 4: Applied DWT on the watermark image.

Step 5: Modified the LL sub-band of the original copy using the LL sub-band of the watermark.

newhost_LL = h_LL + (α*w_LL);

Step 6: Applied inverse DWT and found the watermarked image.

Recovery Process.

Step 1: Selected a watermark image.

Step 2: Applied DWT on the watermarked image.

Step 3: Chose a newhost_LL sub-band.

Step 4: Modified newhost_LL sub-band using the LL sub-band of the watermark image.

newwatermark_LL = (wm_LL-h_LL)/α.

Step 5: Applied inverse DWT and recovered the watermark image.

Algorithm 2: based on Modifying RGB Color frequency component.

Embedding Process.

Step 1: Selected a host image and applied DWT.

Step 2: Chose a low-frequency subband and converted the components of colors red, blue, and green.

Step 3: Selected a watermark image and applied DWT.

Step 4: Chose a low-frequency subband and converted its color components.

Step 5: Embedded the red, blue, and green components of the host image into watermark with factor α.

Step 6: Reconstructed a low-frequency component of the original image by combining new and modified color components.

Step 7: Applied DWT to combine all vertical and horizontal low and high-frequency bands of the host image.

Recovery Process.

Step 1: Selected the watermarked image and apply DWT.

Step 2: Chose a low-frequency band and converted its RGB color component.

Step 3: Selected the original image and applied DWT.

Step 4: Chose low-frequency subband and converted its RGB color component.

Step 5: Reconstructed the RGB color component of the watermark image with the help of the RGB component of watermarked and host images.

New_Watermark_Red = (Red component of watermarked—Red component of host image)/α;

NewWatermarkGreen = (Green component of watermarked—Green component of host image)/α;

New_Watermark_Blue = (Blue_component of watermarked—Blue component of host image)/α;

Step 6: Reconstructed low-frequency subband of watermark image using modified Red, Green, and Blue components.

Step 7: Combined all vertical and horizontal low and high bands of frequency in watermark image using inverse DWT and get the Recovered Watermark image.

2 Combined Approach of DWT and SVD Image Watermarking

Xing and Tan [19] used block-SVD and the Arnold transform to device a watermarking scheme. Experiments confirmed that the block-SVD technique is quicker and more efficient than the normal SVD algorithm. For SVD, the block size was adjusted with the watermark image size. Watermark's pixels are hidden in the initial nonzero unit value bits of each block generated from the block-SVD transformation of the host image. Scrambling with Arnold transform is done for obtaining the security and robustness against cropping image attack. The technique is resistive to the usual attacks such as the addition of Gaussian noise with variance, weak JPEG compression, low pass filtering, rescaling, and cropping.

Sivavenkateswara et al. [34] gave a hybrid watermark mechanism on SVD, DWT, DCT, and Particle Swarm Optimization (PSO). Watermark was implanted on principle constituents of the watermark in the DCT domain of DWT horizontal subbands of the host image. It was done to maintain the imperceptibility, reliability and to avoid the false-positive problems. PSO identifies the optimum scaling factor such that to provide maximum transparency, robustness and imperceptibility.

The Discrete Fourier Transform (DFT) plays an important role in watermarking. It breaks the image as a matrix into its sine components and cosine components. An image in a spatial domain is an input to the DFT, and the outcome is the same image in the frequency domain. Ansari et al. [35] developed a watermarking algorithm that was built on DWT, DFT and SVD. In this approach, the unit values derived from DWT + DFT transformed watermark were entrenched in the color image. SVD can be combined with a lot of other techniques to get better results in image processing. Chung et al. [36] proposed a watermarking scheme that was created using SVD and vector quantization. This method provided a respectable compression ratio and satisfactory image quality.

Lai [37] introduced a watermarking approach, which was developed on SVD and human visual characteristics. It utilized the properties of SVD. It was an efficient representation of the inherent algebraic features of a digital image. The quality and sturdiness of the watermark image are preserved by using HVS characteristics.

Parmar [38] introduced a watermarking scheme built on SVD, DWT, and fuzzy logic. It was a robust scheme against the attacks like salt and pepper and poison attacks.

2.1 Image Watermarking Algorithm based on DWT–SVD

The embedding algorithm.

It is required that the dimensions of the cover object and the watermark must be the same. The process is described as follows:

Step 1: An image is selected.

Step 2: DWT was carried out on the cover image to break down into four sub-bands.

Step 3: Chose the LL1 sub-band and divided its subbands into 8×8 blocks. Every 8×8 size block was then transformed into its RGB component.

Step 4: Applied SVD on each RGB component of the original image.

Step 5: Selected a watermark image.

Step 6: Repeated steps 2–5 for the watermark image.

Step 7: Applied SVD on each RGB element of the watermark (QR Code) image.

Step 8: Embedded the singular values of RGB components of all watermark's blocks into the conforming blocks of cover image with a factor α.

$$sh = sh_1 + \alpha \times sw_1$$

Step 9: New blocks were recreated with the aid of the newly generated *sh*, *u*, and *v* components of every block of the RGB constituent of the original image.

Step 10: The RGB components of each block were then merged.

Step 11: All the blocks of the cover image were merged too.

Step 12: The modified subbands were subjected to inverse DWT, and the watermarked image was generated.

The extraction processes.

Step 1: The watermarked image was called.

Step 2: Applied DWT to disintegrate the watermarked image into four sub-bands.

Step 3: The selected LL1 subband was segregated into 8 × 8 blocks.

Step 4: All the 8 × 8 size blocks were then converted into their respective RGB components.

Step 5: All the RGB components of the watermarked image block were exposed to undergo SVD.

Step 6: Pulled up the watermark image and reiterate the steps 2–5.

Step 7: Applied SVD on every RGB element of the watermark image (QR Code).

Step 8: Selected the raw image and repeated the steps 2–5 on it.

Step 9: Applied SVD on every RGB component of the unprocessed image block.

Step 10: Extracted the singular values of all the RGB components from every watermark block from related blocks of the watermarked image with a factor α.

Step 11: The new blocks were recreated with the aid of fresh sw, u, and v component of every single chunk of RGB constituent of the watermark image.

Step 12: After that, the RGB components of each block were put together.

Step 13: All watermark image blocks were combined too.

Step 14: The watermark image was regenerated by applying inverse DWT with modified subbands.

3 Conclusion

Nowadays, the security of digital content is a challengeable task. There are several approaches, such as cryptography, steganography, and watermarking, to resolve the security of digital content. For ownership and authentication of a digital content image, watermarking is an essential and very interesting research topic. This paper deeply reviewed and explained different watermarking types, techniques, and algorithms. It is based on image processing in the spatial and transformed domain. In spatial domain technique, this paper deeply reviewed least significant bit image watermarking technique and algorithm. Different techniques using singular value decomposition, DCT and DWT in transform domain have been explained. The basic algorithm step of LSB, DCT, DWT and combined approach of DWT-SVD also deeply explained with their result. The main motive of this paper is providing the basic structure of different types of watermarking schemes.

Fig. 1 **a** Original image **b** Watermark Image **c** Watermarked image **d** Recovered Watermark

Fig. 2 **a** Original image **b** Watermark image **c** Watermarked image **d** Recovered watermark image

Fig. 3 **a** Lena **b** Watermark **c** Watermarked **d** Recover watermark

References

1. Li Y-L, Li J-P, Ren Q-B (2010) Based on chaotic encryption and SVD digital image watermarking. In: International Conference on Apperceiving Computing and Intelligence Analysis Proceeding, pp 285–289. IEEE
2. Sun R, Sun H, Yao T (2002) A SVD-and quantization based semi-fragile watermarking technique for image authentication. In: 6th International Conference on Signal Processing, pp 1592–1595. IEEE
3. Agarwal R, Santhanam M, Venugopalan K (2011) Multichannel digital watermarking of color images using SVD. In: International Conference on Image Information Processing, pp 1–6. IEEE
4. Bhuyan T, Srivastava VK, Thakkar F (2016) Shuffled SVD based robust and secure digital image watermarking. In: International Conference on Electrical, Electronics, and Optimization Techniques, pp 1229–1233. IEEE
5. Durga KV, Mamatha G, Bindu CH (2015) SVD based image watermarking with firefly algorithm. In: International Conference on Computer Communication and Informatics, pp 1–7. IEEE
6. Golshan F, Mohammadi K (2011) A hybrid intelligent SVD-based digital image watermarking. In: 21st International Conference on Systems Engineering, pp 137–141. IEEE
7. Kandpal H, Katiyar H, Kumar M (2012) Wave atom-SVD based digital image watermarking scheme. In: Students Conference on Engineering and Systems, pp 1–4. IEEE
8. Liu JJ, Jiang H, Liang GY (2010) A watermarking algorithm for digital image based on logistic and SVD. In: International Conference on Intelligent Computing and Integrated Systems, pp 140–144. IEEE

9. Nana Z (2016) Watermarking algorithm of spatial domain image based on SVD. In: International Conference on Audio, Language and Image Processing, pp 361–365. IEEE
10. Pei S, Liu H, Liu T, Liu K (2010) Color image watermarking using SVD. In: IEEE Conf. on Multimedia and Expo, pp 19–23. IEEE
11. Zhu S, Liu B, Chai Y (2012) An improved self-adaptive digital watermarking algorithm for RGB images based on W-SVD. In: International Conference on Management of e-Commerce and e-Government, pp 57–62. IEEE
12. Zhang L, Tang B (2012) A combination of feature-points-based and SVD-based image watermarking algorithm. In: International Conference on Industrial Control and Electronics Engineering, pp 1092–1095. IEEE
13. Shaw AK, Majumder S (2016) Multiresolution SVD and pixel-wise masking based image watermarking. In: 2nd International Conference on Control, Instrumentation, Energy & Communication, pp 193–196. IEEE
14. Radharani S, Valarmathi M (2014) Content based image watermarking scheme using block SVD and Arnold Transform. In: International Conference on Electronics and Communication Systems, pp 1–4. IEEE
15. Singh RK, Shaw DK, Alam MJ (2015) Experimental studies of LSB watermarking with different noise. Procedia Comput Sci 54:612–620
16. Singh RK, Shaw D (2016) Experimental analysis of impact of noise on various edge detection techniques. In: Proceedings of the World Congress on Engineering
17. Singh RK, Kumar B, Shaw DK, Khan DA (2018) Level by level image compression-encryption algorithm based on quantum chaos map. J King Saud Univ Inf Sci
18. Singh RK, Shaw DK, Sahoo J (2017) A secure and robust block based DWT-SVD image watermarking approach. J Inf Optim Sci 38:911–925
19. Xing Y, Tan J (2007) A color watermarking scheme based on block-SVD and Arnold transformation. In: Second Workshop on Digital Media and its Application in Museum & Heritages, pp 3–8. IEEE
20. Gu Z-F, Li G-F, Yang Z-X (2012) Study on digital image watermark algorithm based on chaos mapping and DWT. In: International Conference on Advanced Mechatronic Systems, pp 160–164. IEEE
21. Guannan Z, Shuxun W, Guijun N (2004) A blind watermarking algorithm based on DWT for color image. In: The 2004 Joint Conference of the 10th Asia-Pacific Conference on Communications and the 5th International Symposium on Multi-Dimensional Mobile Communications Proceeding, pp 634–638. IEEE
22. Banoci V, Bugar G, Levický D (2011) A novel method of image steganography in DWT domain. In: 21st International Conference Radioelektronika, pp 1–4. IEEE
23. Gao H, Zheng X (2010) Image information hiding algorithm based on DWT with alterable parameters. In: International Conference on Information Theory and Information Security, pp 279–282. IEEE
24. Pramoun T, Amornraksa T (2012) Improved image watermarking scheme based on DWT coefficients modification in LL sub-band. In: 9th International Conference on Electrical Engineering/Electronics, Computer, Telecommunications and Information Technology, pp 1–4. IEEE
25. Zhou Y, Jin W (2012) A robust digital image multi-watermarking scheme in the dwt domain. In: International Conference on Systems and Informatics, pp 1851–1854. IEEE
26. Zhongqin W, Lu H, Yang S, Qiang Z, Chaoxia W (2015) Study of digital watermark based on chaos algorithm. In: Third International Conference on Cyberspace Technology. IEEE
27. Wang Q, Ding Q, Zhang Z, Ding L (2008) Digital image encryption research based on dwt and chaos. In: Fourth International Conference on Natural Computation, pp 494–498. IEEE
28. Sun T, Shao X, Wang X (2008) A novel binary image digital watermarking algorithm based on DWT and chaotic encryption. In: The 9th International Conference for Young Computer Scientists, pp 2797–2802. IEEE
29. Liu S, Chen J (2016) A fast multi-focus image fusion algorithm by DWT and focused region decision map. In: Asia-Pacific Signal and Information Processing Association Annual Summit and Conference, pp 1–7. IEEE

30. Thongkor K, Mettripun N, Pramoun T, Amornraksa T (2013) Image watermarking based on DWT coefficients modification for social networking services. In: 10th International Conference on Electrical Engineering/Electronics, Computer, Telecommunications and Information Technology, pp 1–6. IEEE
31. Molina-Garcia, J., Reyes-Reyes, R., Ponomaryov, V., Cruz-Ramos, C.: Watermarking algorithm for authentication and self-recovery of tampered images using dwt. In: 9th International Kharkiv Symposium on Physics and Engineering of Microwaves, Millimeter and Submillimeter Waves, pp 1–4. IEEE
32. rong Wei C, Liu JJ, Liang GY (2010) A DCT_SVD domain watermarking algorithm for digital image based on Moore-model cellular automata scrambling. In: International Conference on Intelligent Computing and Integrated Systems, pp 104–108. IEEE
33. Majumder S, Das TS, Mankar VH, Sarkar SK (2009) SVD and error control coding based digital image watermarking. In: International Conference on Advances in Computing, Control, and Telecommunication Technologies, pp 60–63. IEEE
34. Rao VS, Shekhawat RS, Srivastava V (2012) A DWT-DCT-SVD based digital image watermarking scheme using particle swarm optimization. In: IEEE Students' Conference on Electrical, Electronics and Computer Science, pp 1–4. IEEE
35. Ansari R, Devanalamath M, Manikantan K, Ramachandran S (2012) Robust digital image watermarking algorithm in DWT-DFT-SVD domain for color images. In: IEEE Conference on Communication, Information & Computing Technology, pp 1–6. IEEE
36. Chung K-L, Shen C-H, Chang L-C (2001) A novel SVD-and VQ-based image hiding scheme. Pattern Recognit Lett 22:1051–1058
37. Lai C-C (2011) An improved SVD-based watermarking scheme using human visual characteristics. Opt Commun 284:938–944
38. Parmar A (2015) Image security using watermarking based on DWT-SVD and fuzzy logic. In: 4th International Conference on Reliability, Infocom Technologies and Optimization (ICRITO)(Trends and Future Directions, pp 1–6. IEEE

Image Retrieval Scheme Using Efficient Fusion of Color and Shape Moments

Naushad Varish and Priyanka Singh

Abstract Due to the tremendous increase in the digital image data, the efficient and effective image content-based search scheme for retrieving desired images from a large image repository is highly required. The biggest challenge in image retrieval scheme is to retrieve the desired multimedia images from the digital image repository with minimum time. Extracting significant image features with low dimensional feature descriptor play a significant role in improving retrieval outcomes. In the presented paper, an image retrieval scheme is proposed using fused low dimensional feature descriptor which is obtained by fusion of probability histogram-based HSV color moments and multiresolution based shape moments. The color moments and shape moments are extracted from the Laplacian filter based preprocessed image. The suggested scheme is implemented on a standard Corel-1K image dataset and the retrieval accuracy is measured using precision, recall, and F-score metrics. The experimental outcomes are also validated and compared with some existing state of the art image retrieval schemes and it outperforms over the existing ones.

Keywords Content-based image retrieval · Feature extraction · Statistical moments · Gaussian image pyramid · Normalization

1 Introduction

Content-based Image Retrieval (CBIR) [1, 2] is a scheme of browsing and extracting the required multimedia images from the digital image repository using different visual contents of the image. In CBIR, visual contents represent the color, texture,

N. Varish (✉)
Department of Computer Science and Engineering, Koneru Lakshmaiah Education Foundation, Guntur 522 502, Andhra Pradesh, India
e-mail: naushad.cs88@gmail.com

P. Singh
Department of Computer Science and Engineering, SRM University-AP, Amaravati 522 502, Andhra Pradesh, India
e-mail: priyankasingh401@gmail.com

R. Patgiri et al. (eds.), *Proceedings of International Conference on Big Data, Machine Learning and Applications*, Lecture Notes in Networks and Systems 180,
https://doi.org/10.1007/978-981-33-4788-5_16

shape, and/or any other spatial information of an image [3]. In the last two decades, CBIR has become a proficient image retrieval scheme for extracting the most desired images from the digital repository using automatic extracted visual image contents without any human intervention [4]. Several researchers are trying to design effective and efficient CBIR schemes because nowadays these retrieval schemes are prominently used in a wide range of applications like crime prevention, face recognition, medical imaging, etc. Most of the earlier CBIR schemes have considered only a single visual feature descriptor for retrieving images, but alone feature descriptor is not adequate to identify and distinguish the image visual contents effectively. Since image data consists of many visual structures and shapes, it was difficult to obtain satisfactory retrieval results using only a single visual feature. In order to overcome this limitation, CBIR scheme [5] has been developed with fused visual image descriptor, which is the mixture of low-level visual contents, i.e., color, shape, and texture. Among many visual feature descriptors, the color is mostly used in image retrieval systems due to its simplicity and it is also invariant to rotation, scaling, and any other spatial transformations. Liu et al. [6] have extracted image visual contents based on color difference histograms using specific quantization techniques for effective image retrieval. Many other CBIR schemes have been also developed using visual color feature descriptors like co-occurrence matrix [7], color autocorrelation [8], exponent color moments [9], a color histogram [10], and color moments [11] under a specific color model. Color moments generally represent the distribution of color properties of the image data. These moments are also used to find valuable image contents under various lighting environments and derived from different color spaces like HSV, HSL, YCbCr, RGB color spaces, etc. Verma et al. [12] have developed visual-based image retrieval scheme using local extrema co-occurrence pattern, where these patterns have been determined from three components of HSV color image. Alone color visual features are not adequate to identify and distinguish the image contents effectively, as two different kinds of digital images may have identical histograms. Shape is another visual feature descriptor which represents complex structures and an object in the image effectively. Many shape feature descriptor based image retrieval schemes [13, 14] have been developed, where visual shape contents have been computed using an enhanced version of Zernike moments, shape descriptor based on Generic Fourier transform, shape objects using Canny edge detection, and hue moments, etc., Akrem et al. [15] have developed shape-based image retrieval scheme, where shape signatures have been extracted based on the Fourier descriptor (FD) and farthest point distance (FPD) approach. Kothyari et al. [16] have computed the seven invariant moments from the image for representing shape descriptor. In the presented image retrieval technique, geometric visual shape characteristics are derived from the image which represents the object or shapes of image at different multiresolution level effectively. Since it is difficult to detect objects at one resolution level while other objects may be recognizable at different resolutions effectively. Earlier developed contents-based image retrieval schemes are mainly focused on visual feature descriptors like color, shape signatures, and texture contents alone. Based on the aforesaid discussion, the motivation is to design an improved image retrieval system by mingling color and shape visual contents rather than considering just single content among many. In the presented paper, the aim of the proposed CBIR scheme is to increase the retrieval accuracy using efficient fusion of the color and shape visual contents in a new fashion. The main contribution of the proposed image retrieval scheme is highlighted by the following points:

1. The color visual features have been extracted from the probability histogram-based preprocessed HSV color image.
2. The shape visual features have been extracted from grayscale image using multiresolution-based geometric moments
3. The proficient combination of the extracted color and shape visual contents has been performed in the image retrieval.
4. The Euclidean distance has been used to extract the desired images from the digital repository.

The remaining sections of the paper are organized as follows. Section 2 elaborates the developed image retrieval scheme. The experimental results and discussions are presented in Sect. 3. Finally, the conclusion of paper is provided in Sect. 4.

2 Proposed Image Retrieval Methodology

In general, the image content-based retrieval is done in two main steps; the first step is used to construct the visual feature descriptors of the target images and query image by extracting their visual contents. In the second step, the appropriate similarity distance between the feature descriptors is computed for retrieving the most relevant images from the image repository. Figure 1 depicts the CBIR diagram of the proposed scheme.

2.1 Laplacian Filter Based Preprocessing

In this subsection, initially an RGB input color image is preprocessed using Laplacian filter. Since the human visual system is more responsive to image edges, isolated points and much significant information of an image. In order to find thin edges and various types of isolated points, the Laplacian filter [17] is adapted for computing much finer detailed image, because it is determined by the second-order partial differentiation while other filters/masks like prewitt and sobel are commonly consider only first-order derivatives. The Laplacian mask is convolved with every pixel point of the image. This mask is also pertinent for color images and it covers horizontal and vertical edges effectively. Hence, the differential vector **c** is obtained

$$\nabla^2 [\mathbf{c}(x, y)] = \begin{bmatrix} \nabla^2 R(x, y) \\ \nabla^2 G(x, y) \\ \nabla^2 B(x, y) \end{bmatrix} \tag{1}$$

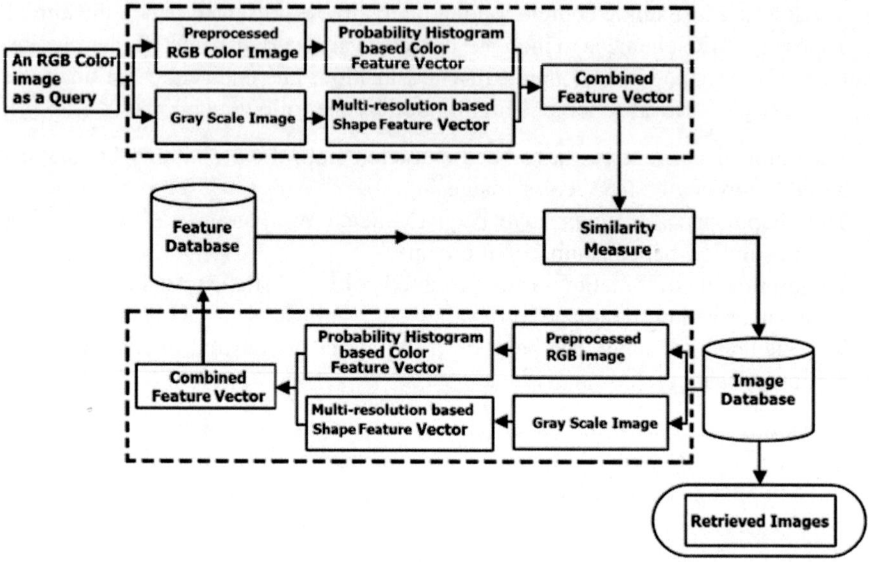

Fig. 1 CBIR diagram of the proposed scheme

The color components, i.e., red(R), green(G), and blue(B) of an RGB color image are differentiated individually; The variables x and y are horizontal and vertical coordinates. Let $I_{cc}(x, y), cc \in \{R, G, B\}$ be an RGB color image, the Laplacian derivate is computed as

$$\nabla^2 [I_{cc}(x, y)] = \frac{\partial^2 I_{cc}(x, y)}{\partial x^2} + \frac{\partial^2 I_{cc}(x, y)}{\partial y^2}$$

where,

$$\frac{\partial^2 I_{cc}(x, y)}{\partial x^2} = I_{cc}(x+1, y) - 2I_{cc}(x, y) + I_{cc}(x-1, y)$$
$$\frac{\partial^2 I_{cc}(x, y)}{\partial y^2} = I_{cc}(x, y+1) - 2I_{cc}(x, y) + I_{cc}(x, y-1) \tag{2}$$

Then,

$$\nabla^2 [I_{cc}(x, y)] = I_{cc}(x+1, y) - 4I_{cc}(x, y) + I_{cc}(x-1, y) + I_{cc}(x, y+1) + I_{cc}(x, y-1)$$

The sharpened/enhanced color image is obtained using a 3D Laplacian mask from the color image in an RGB color space. The color image I initially convolved with a Laplacian mask to obtain an image $L1$ and it has image edges in horizontal and vertical directions. The Laplacian operator is nothing but 3×3 sized matrix with center -4. This mask/matrix is depicted in equation (3). Hence, to determine the enhanced image, the subtraction operation is performed between image $L1$ and image I.

$$\begin{bmatrix} 0 & 1 & 0 \\ 1 & -4 & 1 \\ 0 & 1 & 0 \end{bmatrix} \tag{3}$$

During the image convolution, little bit significant information of the image is lost. To determine this significant information, an enhanced image g is calculated

$$g = I - L_1 \tag{4}$$

The obtained RGB color image g is transformed into color image in a HSV color space. This HSV color image is used to extract color and shape moments for feature representation in image retrieval process.

2.2 *Histogram-Based Color Moments*

Color feature is an extensively used visual content of the image because of its computational simplicity and invariant properties with rotation, scaling, translation, and any other spatial transformation of image [18]. The appropriate global feature representation is used to minimize the computational overhead and increase image retrieval performance to a certain extent. The color moments represent the distribution of colors in the image and it can be computed from color image in any color model. HSV color model is considered in the proposed scheme for color feature extraction. The reason for choosing HSV color model is that it is very close to the human visual perception system. In the presented work, we have taken the preprocessed HSV color image and disintegrated into its hue, saturation, and value components denoted as H, S, and V, respectively. After that, the histograms of each color components have been constructed and corresponding probability histograms have been computed for calculating moments. Statistical/color moments such as mean, standard deviation, skewness, and kurtosis from each probability histogram of color planes of HSV color image is computation for the formation of color feature descriptor. The statistical moments can be calculated as

$$\begin{aligned}
&\text{Mean} && \mu_{CC} = \sum_{r=0}^{L-1} r P(r) \\
&\text{Standard deviation} && \sigma_{CC} = \sqrt{\sum_{r=0}^{L-1} (r - \mu_{CC})^2 P(r)}, \\
&\text{Skewness} && sk_{CC} = \frac{1}{\sigma^3} \sum_{r=0}^{L-1} (r - \mu_{CC})^3 p(r) \\
&\text{Kurtosis} && k_{CC} = \frac{1}{\sigma^4} \sum_{r=0}^{L-1} (r - \mu_{CC})^4 p(r)
\end{aligned}$$

where $CC \in \{H_H, S_H, V_H\}$ and Probability $P(r) = \frac{\text{Number of Pixels at bin } 'r'}{\text{Width} \times \text{Height}}$

(5)

where $CC \in \{H_H, S_H, V_H\}$ represents the probability histogram of color components of HSV color image and pixel values are ranging from $0 - (L - 1)$. Mean represents brightness and the average color information of the image. The contrast of the image is depicted by standard deviation(σ_{CC}). Also, it measures the distribution of pixel values around mean(μ_{CC}) in the histogram. The skewness measures the skewed pixel values of the image about mean in histogram. The kurtosis (k_{CC}) calculates the peakness of the image pixel values. The collective color moments of all three probability histograms of HSV color planes has constructed the 12-D feature descriptor. Hence, color feature descriptor is defined as

$$FV_{\text{Color}} = \{\mu_{CC}, \sigma_{CC}, sk_{CC}, k_{CC}\}, \quad CC \in \{H_H, S_H, V_H\} \tag{6}$$

2.3 *Multiresolution Based Shape Moments*

Shape feature is the most important visual content for identifying the objects and structures in the image [19]. These objects can be easily recognized by human beings using their structures in the image. Major objective of the shape representation is to determine attributes of object; those attributes will be useful in the matching process. In this paper, author has considered a preprocessed image and converted it into a grayscale image for extracting invariant moments as a shape feature. Before computing the moments directly, the grayscale image is analyzed at different resolutions using Gaussian image pyramid approach [20]. Let two dimensional discrete image $I(x, y)$, then moments of order p and q of discrete image is defined as

$$m_{p,q} = \sum_x \sum_y x^p y^q I(x, y), \quad \forall\ p,\ q = 0,\ 1,\ 2$$

where x and y are spatial coordinates of the image. The central moments are obtained as

$$\mu_{pq} = \sum_x \sum_y (x - \overline{x})\,(y - \overline{y})\,I(x, y)$$
$$\overline{x} = m_{10}/m_{00},\ \overline{y} = m_{01}/m_{00}$$

The normalized central moment η_{pq} are

$$\eta_{pq} = \mu_{pq}/\mu^{\gamma}{}_{00},\ \ \gamma = \frac{p+q}{2} + 1,\ \ \forall\, p+q = 2, 3 \ldots$$

The set of seven moments that are invariant with respect to translation, rotation, and scaling are calculated using normalized central moments of the 2nd and 3rd order as follows:

$$\begin{aligned}
\phi_1 &= \eta_{20} + \eta_{02} \\
\phi_2 &= (\eta_{20} + \eta_{02})^2 + (4\eta_{11})^2 \\
\phi_3 &= (\eta_{30} + 3\eta_{12})^2 + (3\eta_{21} - \eta_{03})^2 \\
\phi_4 &= (\eta_{30} + \eta_{12})^2 + (\eta_{21} - \eta_{03})^2 \\
\phi_5 &= (\eta_{30} + 3\eta_{12}) + (\eta_{30} + \eta_{12})\left[(\eta_{30} + \eta_{12})^2 - 3(\eta_{21} + \eta_{03})^2\right] \\
&\quad + (3\eta_{21} + \eta_{03})\,(\eta_{21} + \eta_{03})\left[3(\eta_{30} + \eta_{12})^2 - (\eta_{21} + \eta_{03})^2\right] \\
\phi_6 &= (\eta_{20} - \eta_{02})\left[(\eta_{30} + \eta_{12})^2 - 3(\eta_{21} + \eta_{03})^2\right] + 4\eta_{11}\,(\eta_{30} + \eta_{12})\,(\eta_{21} + \eta_{03}) \\
\phi_7 &= (3\eta_{21} - \eta_{03})\,(\eta_{30} - \eta_{12})\left[(\eta_{30} + \eta_{12})^2 - 3(\eta_{21} + \eta_{03})^2\right] \\
&\quad - (\eta_{30} - 3\eta_{03})\,(\eta_{21} + \eta_{03})\left[3(\eta_{30} + \eta_{12})^2 - (\eta_{21} + \eta_{03})^2\right]
\end{aligned}$$

The set of these seven moments represents the shape feature descriptor of an image and it is given as

$$FV_{mu} = \{\phi_1,\ \phi_2,\ \phi_3,\ \phi_4,\ \phi_5,\ \phi_6,\ \phi_7\} \tag{7}$$

In the proposed shape features extraction process, firstly the grayscale image is decomposed into three different resolutions and from each decomposed sub gray scale image, the set of seven moments have been computed. The collective representation of the set of moments of all decomposed sub images has formed a shape feature descriptor. Then single shape feature descriptor is obtained as

$$FV_{\text{shape}} = \{FV_{mu1},\ FV_{mu2},\ FV_{mu3}\} \tag{8}$$

where FV_{mu1}, FV_{mu2}, and FV_{mu3} represent the feature descriptors of grayscale sub images at 1, 2, and 3 decomposition levels of Gaussian image pyramid, respectively.

2.4 *Fused Features*

In general, natural images are complex and clumsy in nature. These are identified and recognized by its visual contents like color and/or shape feature. To represent the characteristics of both color and shape features simultaneously, a specific fusion technique is adapted in the proposed scheme. The feature extraction block diagram of feature fusion is depicted in Fig. 2. The feature components should be normalized to make the features in the same range/scale of components and to achieve the unit feature representation. Different multimedia content may have different range/scale of components. Thus, normalization of feature components is inevitable to avert the components variances while measuring the similarity. The single visual feature descriptor is obtained by combining the color and shape features as follows:

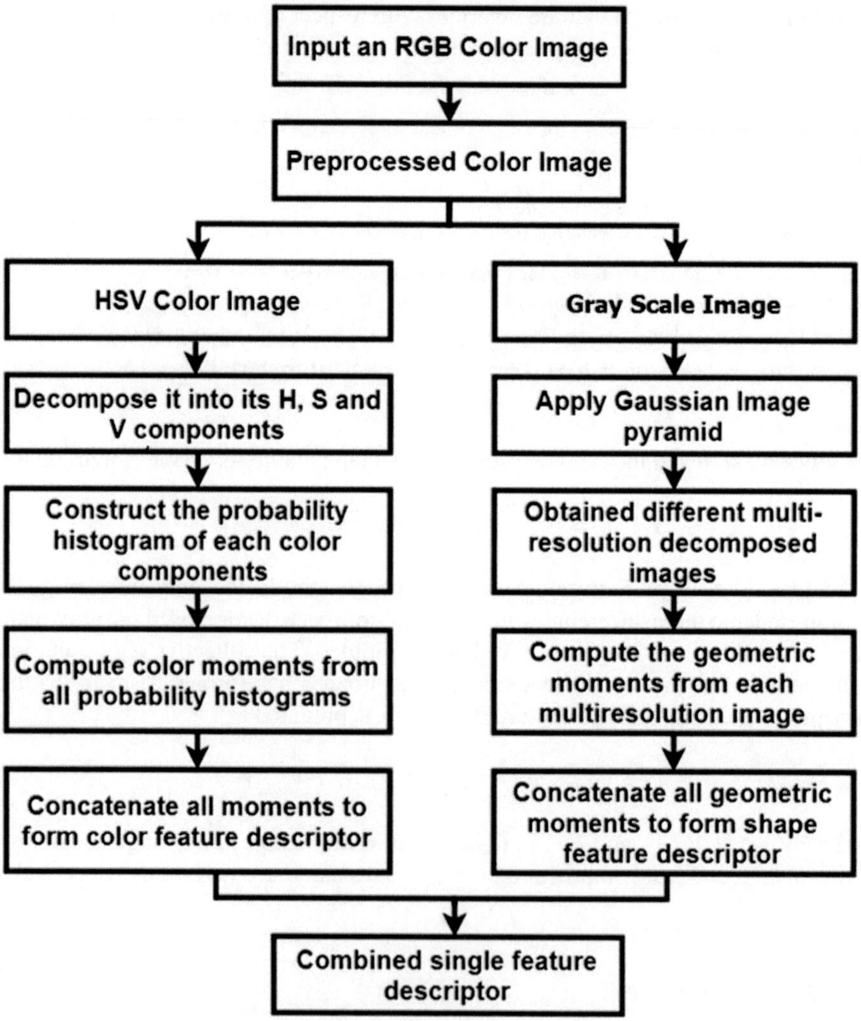

Fig. 2 Block diagram of proposed feature extraction

$$FV = \left[\frac{(FV_{\text{Color}} - \mu_{\text{Color}})}{\sigma_{\text{Color}}}, \frac{(FV_{\text{shape}} - \mu_{\text{shape}})}{\sigma_{\text{shape}}} \right] \tag{9}$$

where μ_{Color} and μ_{shape} represent the means, σ_{Color} and σ_{shape} are standard deviations of the color and shape feature descriptors, respectively. The color and texture features are normalized using z-score normalization technique. The obtained single feature descriptor follows the Gaussian shape structure. This feature descriptor will be used for retrieving the images with highest similarity from the repository.

2.5 *Similarity Measure and Evaluation Metrics*

The performance of image retrieval scheme relies on the appropriate extracted image contents and similarity measures. In the proposed scheme, Euclidean distance is considered for measuring the similarity between a query image and database images. The Euclidean distance is square root of the sum of the squares of differences of components of feature descriptors of query image and target images in the repository. It is defined as

$$D = \sum_{i=1}^{d} \sqrt{|FV_i(q) - FV_i(t)|^2}, \quad \forall \; i = 1, 2, \ldots, d \tag{10}$$

where $FV(q)$ is the feature descriptors of the query image and $FV(t)$ represent target image in the repository; d denotes length of feature descriptor. Two images are considered as similar if $d = 0$. The accuracy of the any image retrieval system is commonly measured in terms of precision and recall metrics based on a given query image. The highest value of precision indicates the best retrieval results, while the lowest value shows worst results. The recall is opposed to the precision; the highest value of recall indicates the worst results and lowest value provides the best retrieval results. We can define precision(P) and recall(R) as follows:

$$\begin{aligned} P &= \frac{\text{No. of relevant images retrieved}}{\text{Total no. of images retrieved}} \times 100\% \\ R &= \frac{\text{No. of relevant images retrieved}}{\text{Total no. of relevant images in each category of the database}} \times 100\% \end{aligned} \tag{11}$$

These two metrics are inadequate to give the overall performance of image retrieval system effectively. To compute overall performance, harmonic mean of P and R is calculated and is known as F-measure or F-score. The mathematical formula for F-score is obtained as

$$F = \frac{2 \times P \times R}{P + R} \tag{12}$$

3 Experimental Results and Discussions

The implementation of the proposed CBIR scheme has been done using MatLab R2013a on the platform Intel core i3 CPU @ 2.27 GHz, 6 GB RAM, and windows 7(32-bit operating system). The retrieval accuracy of the proposed scheme is validated on a Corel-1K [21] benchmark image dataset. This dataset has 1000 images of various structures and different contents. Corel-1K dataset categories into ten clusters where each cluster contains one hundred similar JPEG format images with size 256×384 or 384×256. Sample images of Corel-1K dataset is shown in Fig. 3. In the presented work, initially an RGB color is preprocessed using Laplacian filter and converted into HSV color image and grayscale image individually. For color feature extraction, authors have decomposed HSV color image into its hue, saturation, and value components and corresponding probability histograms have been computed. Subsequently, four color moments from each histogram are calculated which represents the color feature descriptor. Dimension of color feature descriptor is $4 \times 3 = 12$. Thereafter, the grayscale image is analyzed at multiresolutions using three level Gaussian image pyramid and the seven moments from each leveled image have been computed for representing shape feature descriptor. In this way, the length of the feature descriptor will be $3 \times 7 = 21$. For representing mixed information of color and shape descriptors, a specific fusion approach is considered which fused these two features in a normalized fashion so that the variation among the feature components should be minimized. The dimension of the single mixed feature descriptors will be $12 + 21 = 33$, which is very low in comparison to the original image size. This lowers the computational overhead. In the experimental results, we have retrieved the top twenty images from the image repository having minimum distances. As discussed earlier, precision(P), recall(R), and F-score(F) percentages are used to find the retrieval performance of the proposed image retrieval scheme. For Corel-1K image dataset, the experimental results are performed with and without preprocessing techniques on the images and the retrieval accuracy in terms of precision, recall, and F-score (in percentages) is depicted in Table 1. In this table, average of P, average of R, and average of F are 71.50, 14.30, and 23.83% for the top 20 retrieved images with preprocessing technique, while these metrics are 56.50, 11.30, and 18.83% without preprocessing. Here, we have found a huge improvement in the

Fig. 3 Sample images of the Corel-1K dataset

Table 1 Precision, recall and F-score for top $L = 20$ retrieved images

SI no.	Category	Without preprocessing			With preprocessing		
		Precision	Recall	F-Score	Precision	Recall	F-Score
1	People	45.00	9.00	15.00	60.00	12.00	20.00
2	Beaches	45.00	9.00	15.00	55.00	11.00	18.33
3	Buildings	40.00	8.00	13.33	50.00	10.00	16.67
4	Buses	75.00	15.00	25.00	95.00	19.00	31.67
5	Dinosaurs	65.00	13.00	21.67	100.00	20.00	33.33
6	Elephants	40.00	8.00	13.33	45.00	9.00	15.00
7	Flowers	95.00	19.00	31.67	70.00	14.00	23.33
8	Horses	55.00	11.00	18.33	95.00	19.00	31.67
9	Mountains	65.00	13.00	21.67	80.00	16.00	26.67
10	Foods	40.00	8.00	13.33	65.00	13.00	21.67
Average		**56.50**	**11.30**	**18.83**	**71.50**	**14.30**	**23.83**

retrieval accuracy in terms of P, R, and F. It is quite evident that the images captured in nature contain many unwanted information that should be removed using some suitable preprocessing technique. To reduce the unwanted information, the image is filtered out by the Laplacian filter which makes the zeros in constant area of an image due to derivative operator. The lowest precision, i.e., 40.00% for buildings, elephants, and foods images have been achieved, while it is highest, i.e., 95.00% for flower images without preprocessing. Some building images may be mixed with beach images making it difficult for the proposed scheme to distinguish the similar kinds of images between two category images. With preprocessing, the elephant images have got the lowest precision while the dinosaur images have got the highest precision. Remaining category images have provided acceptable retrieval results.

The proposed image retrieval scheme is compared with four state-of-the-art CBIR schemes. Comparative study shows that the proposed scheme outperforms the existing ones. Table 2 shows the comparative results, where we found that the proposed scheme got 71.50% precision while others got less precisions. Yildizer et al. [22] have suggested clustering algorithm-based CBIR scheme, where food images have achieved the worst precision while the highest precision for the flower images have been obtained, i.e., 99.50%. Kundu et al. [23] have developed an image retrieval scheme based on multi-scale geometric analysis of non-subsampled contourlet transform. They improved the retrieval accuracy and reduced the semantic gap by integrating a relevance feedback mechanism. The lowest precision for beach images have been recorded and the highest precision for elephant images have been found. Mistry et al. [24] have used the spatial domain features for retrieving images effectively from the database. In this scheme, mountains images have got the worst precision since some contents of these images are a little bit similar to the beach, elephant, and building images. The dinosaur and flower images have achieved the highest precision, i.e., 89.00%. Rashno et al. [25] have extracted the wavelet and color- based image features

Table 2 Comparative results in terms of precision in percentages for top $L = 20$ retrieved images

Category	Yildizer et al. [22]	Kundu et al. [23]	Mistry et al. [24]	Rashno et al. [25]	Proposed scheme
People	50.20	44.00	62.00	58.30	60.00
Beaches	71.10	32.00	71.00	48.90	55.00
Buildings	36.40	52.00	53.00	45.50	50.00
Buses	80.10	62.00	77.00	54.60	95.00
Dinosaurs	92.60	40.00	89.00	99.80	100.00
Elephants	60.00	80.00	55.00	46.90	45.00
Flowers	99.50	57.00	89.00	89.60	70.00
Horses	82.60	75.00	55.00	61.20	95.00
Mountains	55.00	57.00	52.00	39.80	80.00
Foods	25.70	56.00	55.00	63.30	65.00
Average	**65.30**	**55.50**	**65.80**	**60.79**	**71.50**

to perform the image retrieval task. In this work, mountain images has got the lowest retrieval precision while the dinosaur images have obtained the highest precision. In the proposed scheme, in almost all category, images have reported the good precision except the elephant images, but the overall performance is acceptable in the present scenario. We have observed a huge improvement in the precision as compared to the existing schemes. In the simulation results, for dinosaur and Elephant images, the top $L = 20$ retrieved images from the Corel-1K image repository using with and without preprocessing techniques shown in Fig. 4, where query image is in the upper top left corner. The Dinosaur image category provides the best precision with the preprocessing technique. The elephant images give the worst results using with and without preprocessing techniques.

4 Conclusions

In this paper, a visual content-based image retrieval scheme using proficient combination of color and shape moments is proposed, where these moments are extracted from the enhanced preprocessed image. The color moments have been computed from the probability histogram of the enhanced HSV color image and shape moments are computed from the multi-level decomposed preprocessed grayscale image. The color moments represent the color feature descriptor, while the shape moments represent the color feature descriptor of the image. The low dimensional single feature descriptor is obtained by fusing these feature descriptors after normalizing individually. The performance metrics, i.e., average precision, average recall, and average F-score of presented image retrieval scheme are adequate. In the comparison, the scheme outperforms other existing schemes in terms of retrieval accuracy.

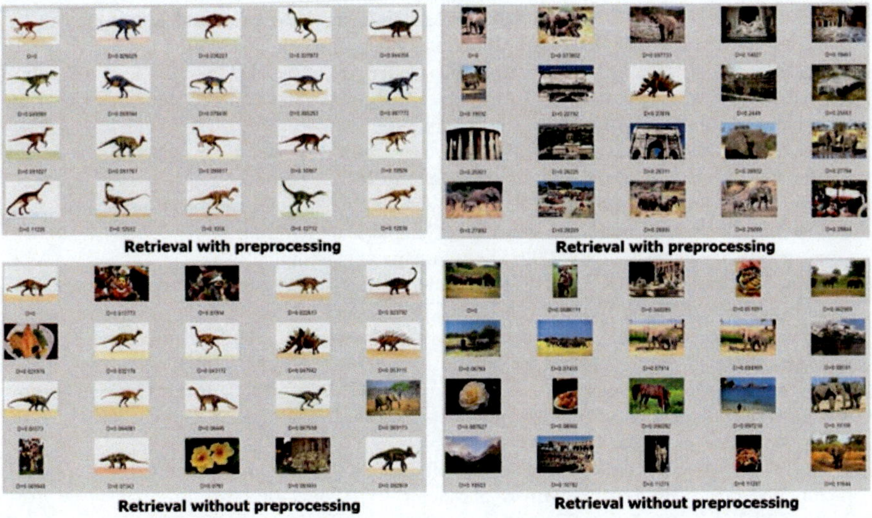

Fig. 4 The retrieval results for dinosaur and elephant images of Corel-1K database with preprocessing and without preprocessing, where queries are the upper top left corner images

References

1. Subrahmanyam M, Jonathan Wu QM, Maheshwari RP, Balasubramanian R (2013) Modified color motif co-occurrence matrix for image indexing and retrieval. Comput Electr Eng 39(3):762–774
2. Jeena Jacob I, Srinivasagan KG, Jayapriya K (2014) Local oppugnant color texture pattern for image retrieval system. Pattern Recognit Lett 42:72–78
3. López-Iñesta E, Grimaldo F, Arevalillo-Herráez M (2017) Combining feature extraction and expansion to improve classification based similarity learning. Pattern Recognit Lett 93:95–103
4. Papushoy A, Bors AG (2015) Image retrieval based on query by saliency content. Digital Signal Process 36:156–173
5. Walia E, Pal A (2014) Fusion framework for effective color image retrieval. J Vis Commun Image Representation 25(6):1335–1348
6. Liu G-H, Yang J-Y (2013) Content-based image retrieval using color difference histogram. Pattern Recognit 46(1):188–198
7. Lin C-H, Chen R-T, Chan Y-K (2009) A smart content-based image retrieval system based on color and texture feature. Image Vis Comput 27(6):658–665
8. Rahman MM, You D, Simpson MS, Antani SK, Demner-Fushman D, Thoma GR (2013) Multimodal biomedical image retrieval using hierarchical classification and modality fusion. Int J Multimedia Inf Retrieval 2(3):159–173
9. Yang H-Y, Na X, Li W-Y, Li Y-W, Niu P, Wang X-Y (2015) Color image representation using invariant exponent moments. Comput Electr Eng 46:273–287
10. Varish N, Pal AK (2015) Content based image retrieval using statistical features of color histogram. In: 2015 3rd international conference on signal processing, communication and networking (ICSCN). IEEE, pp 1–6
11. Bella MIT, Vasuki A (2019) An efficient image retrieval framework using fused information feature. Comput Electr Eng 75:46–60
12. Verma M, Raman B, Murala S (2015) Local extrema co-occurrence pattern for color and texture image retrieval. Neurocomputing 165:255–269

13. Huang M, Shu H, Ma Y, Gong Q (2015) Content-based image retrieval technology using multi-feature fusion. Optik-Int J Light Electron Optics 126(19):2144–2148
14. Imran M, Hashim R, Elaiza N (2015) Content based image retrieval using color layout descriptor and generic fourier descriptor. In: Advanced computer and communication engineering technology. Springer, pp 163–170
15. El-ghazal A, Basir O, Belkasim S (2009) Farthest point distance: a new shape signature for fourier descriptors. Signal Process: Image Commun 24(7):572–586
16. Kothyari P, Dwivedi S, Mandoria HL (2016) Content based image retrieval using statistical feature and shape extraction. Int J Innov Res Comput Commun Eng 4(6)
17. Gonzalez RC, Woods RE et al (2002) Digital image processing
18. Pavithra LK, Sree Sharmila T (2018) An efficient framework for image retrieval using color, texture and edge features. Comput Electr Eng 70:580–593
19. Khokher A, Talwar R (2017) A fast and effective image retrieval scheme using color-, texture-, and shape-based histograms. Multimedia Tools Appl 76(20):21787–21809
20. Yadav AR, Anand RS, Dewal ML, Gupta S (2015) Gaussian image pyramid based texture features for classification of microscopic images of hardwood species. Optik 126(24):5570–5578
21. Wang JZ, Li J, Wiederhold G (2001) Simplicity: semantics-sensitive integrated matching for picture libraries. IEEE Trans Pattern Anal Mach Intell (9):947–963
22. Yildizer E, Balci AM, Jarada TN, Alhajj R (2012) Integrating wavelets with clustering and indexing for effective content-based image retrieval. Knowl-Based Syst 31:55–66
23. Kundu MK, Chowdhury M, Bulò SR (2015) A graph-based relevance feedback mechanism in content-based image retrieval. Knowl-Based Syst 73:254–264
24. Mistry Y, Ingole DT, Ingole MD (2017) Content based image retrieval using hybrid features and various distance metric. J Electr Syst Inf Technol
25. Rashno A, Rashno E (2019) Content-based image retrieval system with most relevant features among wavelet and color features. arXiv preprint arXiv:1902.02059

Detection of Unattended Luggage in Public Places: A Review

C. V. Bijitha

Abstract Security is one of the biggest concerns in present-day life. Attackers or terrorists have always chosen public places for their offences to maximise impact. Thus, surveillance systems are expected to alert even the presence of an unattended item in public places. In this paper, we study various methodologies put forward by researchers over the past decade to solve the abandoned luggage detection problem. Subproblems involved are identified, and the existing approaches used for static object detection, classification of static objects such as bag or non-bag items, and finally owner identification are discussed. Publicly available datasets like PETS2006, iLiDS for AVSS2007, and ABODA for testing abandoned luggage algorithms are familiarised.

Keywords Stationary object · Abandoned object · Unattended luggage detection · Survey · Video surveillance

1 Introduction

Law enforcement officials nowadays undoubtedly depend on video surveillance to fight crime, prevent terrorism, and thus reinforce civic safety [1]. Surveillance systems functioning all over the world collect far more video data than what is being viewed or analysed. An automated approach is required to fulfil the whole purpose of security cameras, which could detect suspicious events in public places, without the need for operator supervision. Here we investigate the state-of-the-art over a decade, in the arena of abandoned object detection. Those methods aim to alert the officials when people leave any object (precisely luggage) in public places that could potentially be a bomb for explosive attacks.

According to [2], abandoned luggage may be defined as baggage that is left by the owners and remain unattended. It is natural that people keep their heavy bags on

C. V. Bijitha (✉)
Kannur, India
e-mail: cvbijitha@gmail.com

R. Patgiri et al. (eds.), *Proceedings of International Conference on Big Data, Machine Learning and Applications*, Lecture Notes in Networks and Systems 180,
https://doi.org/10.1007/978-981-33-4788-5_17

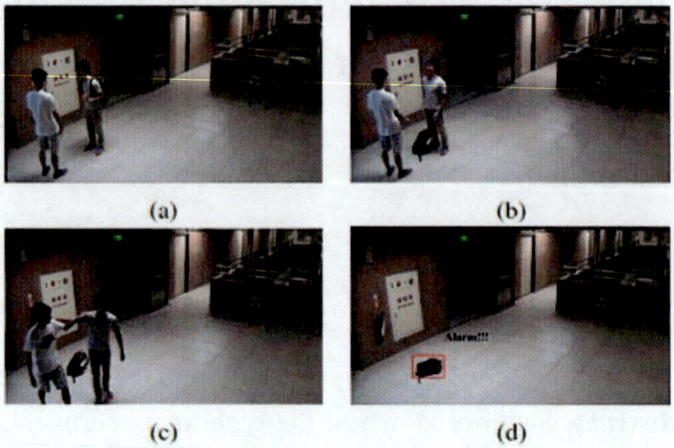

Fig. 1 Abandoned luggage detection

the ground for a while, which cannot be treated as a doubtful event. When there is no physical contact between luggage and the owner (person who brought luggage into the scene), and if he is further away from a specific distance and has not come back and re-attended the baggage in a definite period, an alarm should be triggered and the event should be declared as suspicious.

Occurrence of the following sequence of events in order, decides whether the luggage is abandoned or not, as depicted in Fig. 1.

(1) A person enters with a luggage, we would refer to him as the owner (Fig. 1a, b).
(2) Owner leaves the scene without the bag (Fig. 1c).
(3) Alarm event (Fig. 1d).
(4) Owner may or may not return back to the baggage.

Every frame consists of a static background and moving foreground pixels. Luggage is part of the moving foreground when it is introduced into the scene. When the baggage is left unattended by the owner, the status of those pixels change from moving to static and remains stationary over several frames. And those pixels are called as static foreground [3] in the literature. The main challenge in the problem is to successfully find the static foreground region which are the potential candidates to be an abandoned object/luggage.

The remainder of the paper is organised as follows: Section 2 gives the overall idea of the methods reviewed. Most of the frameworks proceed in a similar pattern. The initial step of identifying the static foreground region is discussed in Sect. 2.1. To avoid false positives, it is essential to confirm whether the static foreground regions are baggage or not. Classification or clustering strategies used for the purpose are discussed in Sect. 2.2. Section 2.3 talks about the methods adopted to find the owner of the left luggage. Works based on deep learning techniques are discussed in Sect. 2.4. Section 3 explore the publicly available datasets for the testing of abandoned luggage detection methods. The effectiveness of each of the methods are critically evaluated in Sect. 4 and finally concluded in Sect. 5.

2 Research in Abandoned Luggage Detection

Abandoned object detection method will become extremely difficult in a crowded environment, scenes with sudden illumination changes, partially or fully occluded objects, etc. All these challenges made the researchers to come up with solutions to detect abandoned objects that could create a threat to the public property and life. This section attempts to review those methods discussed in [4, 6, 8, 9, 11, 13, 15, 17, 19, 23, 24].

2.1 Static Foreground Detection

Porikli et al. [4] attempts to find out the stationary foreground objects by assimilating results from two different background models with Bayesian update as discussed in [5]. Learning rate is an important parameter when we address background subtraction methods. It is evident that initialising learning rate with different values will result in different background models. A background model is called a short-term background model, which learns at a lower scale and is updated at a quicker rate. And the high learning rate yields a model that is updated at a slower pace called the long-term background model. Using both of these models together gives a promising result in detecting stationary objects. Whenever an object becomes stationary, the short-term model which learns at a faster rate incorporates it into the background, but the long-term model will treat it as the foreground itself. Then those pixels are considered to belong to the abandoned item.

Bhargava et al. [6] uses the background initialisation and updating method discussed by Chen and Aggarwal [7]. The approach estimates the background model from the image sequence, in place of choosing the first frame as the background. Then proceeds by identifying the intervals in which the intensity values remain almost constant. The scheme determines which interim is most probable to depict the actual background with the assistance of optical flow data. Once the model is successfully estimated, background subtraction is done to identify the foreground blobs in the HSV colour space. The advantage of choosing an HSV colour space is that it guarantees illumination invariant output. The results then undergo a series of morphological operation which further identifies the candidate blobs of interest.

Chang et al. [8] introduces a foreground-mask sampling to locate the abandoned luggage. Several backgrounds subtracted frames are taken, and their intersection is calculated. Since abandoned objects are part of the static foreground, they will obviously appear in the intersection. The model for the background is built by selecting several cleaner frames (frames with minimum background clutter) and finding its average. Background subtraction is done on n samples and corresponding n foreground images are produced. Then merged n foreground masks' intersection is taken as the static mask. Connected component analysis, along with filtering, is subsequently completed. Regions that persisted in the foreground for n frames over time is a possible abandoned object or stationary human.

Tian et al. [9] uses the mixture of three Gaussian method extended from [10] for the static object detection. The first Gaussian represents background frame, third Gaussian models pixel with quick changes. The second Gaussian is used to detect the static region. To improve the performance, the framework also considers contextual knowledge of the stationary blob to assert the candidate abandoned object.

Fan and Pankati [11] uses the specialised version mixture of 3 Gaussian [12] for the static object detection as in [9]. The second Gaussian represents the static foreground blob which they called as Temporarily Static Objects (TSO). Different states of the TSOs are modelled using a finite state machine.

Chang et al. [13] discusses a unique method to detect abandoned luggage, and their owners using a human classifier learned online [14]. Frames are divided into blocks, and the associated classifier categorises the block as background or foreground. Background patches are considered to be positive samples and foreground patches-the negative samples are sent to train object classifier. Static foreground mask is created by finding the difference between the current and previous frames. After thresholding, obtained static foreground objects and the static foreground mask are passed to the classifier to identify its type. If the static object is a person, it matches with high confidence to those object classifiers. A person with luggage is matched with low confidence.

Kim et al. [15] considers the unattended objects as a new background and the structural changes in the scene are modelled using edge segments. Edge segment distributions are used scene analysis and segmenting the scene to the background, moving foreground and static objects as in [16].

Lin et al. [17] uses both short and long-term background modelling method as in [4] for static object detection. They have used the Mixture of Gaussian method in [18] as their background modelling method. If foreground pixels are represented with a code 1 and background pixels with a code 0, the concatenated result from the long and short model, respectively, gives 10, which indicates the pixel to be a static foreground pixel. On considering a set of frames together, if the state of the same pixel continues to be in 10, it is claimed that the pixel is part of the static foreground. It is modelled using a pixel-based finite state machine (PFSM).

Static foreground detection approach suggested by Wahyono et al. [19] is also based on two background models. It uses the initial frame as the reference model (assuming no static objects in the frame). The second model is built from a set of consecutive frames. The intensity difference of corresponding pixels from the set of frames should be low enough if they are part of the background. Their average value is used to update the background; other pixels are considered as foreground pixels. Dual Background Difference, the difference between the reference model and the updated model is likely to represent the static foreground object. From a sequence of such dual backgrounds, the decision of classifying the pixels as background and background-to-static pixels is taken with the help of K-means clustering. The connected component analysis is performed. Intersections of regions of the successive binary images are taken to confirm the stability and to remove the noise pixels, regions with area more than a threshold are considered.

Cuevas et al. [20] introduced three non-parametric kernel density estimation (KDE) detectors, in which the first one detects moving foreground objects, another one detects recently stopped foreground objects, and the third one detects static foreground objects. Ilias et al. [21] uses the method proposed by [22] to separate moving edges from background. Sobel operator is applied independently in both x and y directions, and the gradient of the background scene is estimated. The background is updated using the difference between the input gradient and background gradient and the foreground edges mask is formed. Temporal accumulation method is used to detect the static edges, and each edge segment is enclosed in a rectangle.

2.2 Classification

Once the static foreground is identified, the next step is to classify the foreground blob as bag or non-bag. The blob is considered to be a large, compact entity that does not normally exceed half the average adult's height. This section discusses the approaches adopted by the researchers to solve the classification problem.

Bhargava et al. [6] have used a simple 3-nearest neighbour algorithm to classify binary foreground blobs applying shape and size as features. The correctness of the classification is ensured by tracking the blob over several continuous frames. If it conforms to the position and there is no other blob in the vicinity, the baggage is reported to be abandoned.

Chang et al. [8] uses skin colour information and body contours to differentiate between human and abandoned luggage from the static foreground blob. The regions identified as humans are discarded, and the candidate luggage objects are retained. Since they are not attempted to learn and classify the blobs as bag or non-bag, they are devoid of training data and associated constraints. A human detector is employed by Tian et al. [9] to classify static blob to a person or other objects. The detector based on adaptive local features is capable of detecting people in the near, mid, or far-field views.

Fan and Pankanti [11] does not explicitly use any classification mechanisms, instead uses the region growing method, which takes into account the similarities between the static regions and its neighbourhood as discussed in [9]. Chang et al. [13] with the aid of online classifiers calculates for each patch a static score. This has a higher value if the patch has lower confidence (a person with luggage is matched with low confidence). Threshold and connected component analysis are performed, and the static foreground is identified.

In [15], moving edges are accumulated into the so-called temporal edge weight map (TEWM). From the incoming frames, this TEWM is analysed, and unattended objects are identified using clustering. Lin et al. [17] collects all the pixels accepted by the PFSM, and connected component analysis is performed, small components are removed and the retained components are considered as an abandoned object. No further classification of retained components to bag or no bag is discussed.

Wahyono et al. [19] obtains feature descriptors for classification from methods based on image moments and geometric properties. The features are trained with the help of SVM [28] and random forest [29] to confirm the decision on baggage and non-baggage. In [20], output of the detectors is passed to the finite state machine for classification. In [21], static foreground is found in such a way that classification of a true or false abandoned object is done by checking for an object in the bounding box.

2.3 *Back-Tracing for Owner Detection*

This section discusses methodologies adopted to find the actual owner of the abandoned baggage.

While unattended baggage is located, Bhargava et al. [6] uses the following approach. The programme tracks it through previous frames and recognises when the luggage was first brought on the scene. The appearance model of all the persons in the scene is captured, and a patch bank is formed. Candidate owners are identified by cross-correlating the foreground blobs in the neighbourhood of the abandoned luggage with the patch bank for a maximum match.

Chang et al. [8] follows the procedure that the system goes back to a time when the owner was still attending the luggage if the owner is not around. As mentioned above, the human body contour and skin colour are used to track the person from that point.

Tian et al. [9] do the owner identification by tracing all the moving objects in the view, and respective trajectories are illustrated. Then the framework identifies the trajectory associated with the detected abandoned object.

In Chang et al. [13], if the static object is a person, high confidence is expected. Once a static object is identified, it is sent to object classifiers and confidence is calculated. The average confidence from several classifiers are calculated, and the owner is identified based on the classifier with the highest confidence.

Lin et al. [17] checks for any other object moving around, if none is found, then it goes back to the time in which the owner brought the bag to the scene. This time is obtained from the PFSM (the time duration in which the pixels stayed in the '10' state). Then, inspect for any human-shaped blobs in a window around the object using a human detector. The foreground region containing the human closest to the object blob is interpreted as the owner and is followed up further. Colour distribution of the foreground blob is utilised for tracking.

To keep the framework simple, Wahyono et al. [19] have not implemented any mechanism to detect the owner of the luggage, stationary baggage was declared to be unattended if there was no non-luggage object at a distance larger than a threshold over the period. Owner identification techniques are not discussed in [4, 11, 15, 20, 21].

2.4 Deep Learning Approaches to Abandoned Luggage Detection

Recently, the thrust for accuracy has shifted the problem-solving methodologies from classical machine learning to deep learning. This section familiarises a few such attempts of solving the abandoned luggage detection problem using deep learning.

Smeureanu and Ionescu [23] used background subtraction and motion estimation to find static objects. The static objects were classified as abandoned using a cascade of two deep convolutional networks. The first of these is used to classify the static objects detected in the earlier steps as luggage or not. The second network of the cascade is used to conclude whether the luggage detected by the previous network is attended by a human or not. Smeureanu and Ionescu [24] detected luggage using background subtraction by Gaussian mixture model followed by luggage recognition using a lightweight convolutional neural network. The method used a two thread implementation with one thread running background subtraction and the second thread running the luggage detection, with real-time performance.

2.5 Other Methods

The algorithms that do not follow the above framework are discussed briefly in this section. A neural network-based approach was introduced by Maddalena and Petrosino in [25] for static object detection. A 3D neural model that automatically reflects and adapts to the changes in the image sequences are implemented. The algorithms that do not follow the above framework are discussed briefly in this section. Meanwhile, Spatio-temporal relationship between a moving person and unattended objects are determined by [26], and a vector matching scheme is applied for detection.

3 Datasets

Researchers have mainly evaluated their frameworks in PETS2006 [2], iLiDS for AVSS2007 [27]. A new dataset called ABODA constructed by [17] is also available in the public domain.

3.1 PETS 2006

The PETS2006 (Performance Evaluation of Tracking and Surveillance) consists of 7 datasets each with 4 video sequences (768 x 576 pixels, 25 fps) recorded with multiple cameras from the Victoria Station in London, UK.

The scenario of a person waiting for a train temporarily placing his briefcase on the ground and then picking it up again is contained in dataset S3. This is not an abandoned luggage event. Apart from this, all other datasets contain sequences involving abandoned luggage scenarios.

3.2 *ILIDS for AVSS2007*

The dataset is taken from i-LIDS (Imagery Library for Intelligent Detection Systems). This dataset is composed of frames with unattended bags on an underground station platform. There are three videos classified based on the complexity. In AVSS AB Easy, luggage is larger in size and the scene is less crowded. In AVSS AB Medium and AVSS AB Hard, luggage size is small, scene is more crowded, consequently more difficult.

3.3 *ABODA*

Lin et al. [17] have constructed ABandoned Objects DAtaset (ABODA), which is publically available. It includes 11 videos shot at different real and challenging scenarios. The situations include crowded scenes, sequences with sudden changes in lighting, sequence recorded at night with indoor and outdoor environments.

Other datasets used by authors include LASIESTA3 [30], ChangeDetection [31], PETS2007 [32], etc.

4 Performance Comparison

Performance of the methods is evaluated mainly with Precision, Recall and F-measure. Where

$$\text{Precision}(P) = \frac{T_P}{F_P + T_P} \quad \text{and} \quad \text{Recall}(R) = \frac{T_P}{F_N + T_P}$$

where T_P is true positives (Abandoned objects are correctly detected), F_P is false positives, T_N is true negatives and F_N is false negatives.

F-measure is the harmonic mean of Precision and Recall.

$$F - \text{Measure} = \frac{2 * R * P}{R + P}$$

Table 1 shows the consolidated results from the discussed methods comparing algorithms performance with PETS2006, AVSS2007 and ABODA.

4.1 Discussion

Like any other image processing/computer vision problem, abandoned luggage detection also suffers from challenges including sudden illumination changes, crowded environment, noise, object shadows, occlusions, etc. Since the methods discussed above are mostly based on background subtraction, it is crucial to get frames for the background initialisation with the least background clutter and containing only static objects. Sudden illumination changes will completely alter the background model and cause the methodology to fail. This scenario is included in the ABODA dataset and is tested by [17] and [19]. If the lighting change is temporary, the use of adaptive background model as in [9] could offer a good solution. Edge-based methods like [15] and [21] could perform better in scenes with gradual lighting change.

Occlusions in the scene cause improper learning of background, and false positives will be generated if the occlusions occur at times when the object is dropped by the owner. Occlusions also cause problems in tracking the owner, where the track is lost due to occlusions as figured out in [8]. Use of slow learning rates as discussed in [11] will less affect the foreground even if the abandoned object is partially occluded. The work in [20] also gives good results in case of occlusions.

Persons stationary over a while are detected as an abandoned item in [4] and [9] since they do not incorporate any object detectors or classifiers. The explicit use of the learning-based method in [13] could discriminate abandoned objects from static persons successfully. When a person sits near the luggage or moves closer to the baggage, they will be blended to the same blob which causes false alarm in [15] and failure of kNN classifier in [6]. Summary of all the discussed methods is given in Table 1.

5 Conclusion

In this paper, we have examined various algorithms proposed for detecting abandoned luggage in public places. The solutions to static object identification, classification and owner identification problems are compared. The above discussed methods, based on background subtraction are very well suited to detect objects in a sequence captured from static cameras. Getting a neat background model from a moving camera is comparatively difficult. False positives of detecting a static human as abandoned objects need to be avoided with the help of strong person detection techniques.

Table 1 Summary

Approach	Datasets used for testing	Results
Porikli et al. [4]	PETS2006, i-LIDS, Own dataset ATC	Person stationary over a period of time is classified as abandoned object
Bhargava et al. [6]	PETS2006, i-LIDS	System fails in 2 test sequences of iLiDS
Liao et al. [8]	PETS2006, i-LIDS	No false alarm are generated
Tian et al. [9]	PETS2006, i-LIDS	Accurate detection of abandoned objects with 1 false alarm each in AB Medium and AB Hard
Fan et al. [11]	i-LIDS, Own Dataset CS10	Static persons and quick lighting changes caused false alarms. Achieved precision of 0.4 and recall of 0.67
Chang et al. [13]	PETS2006, i-LIDS	Detected all abandoned events without any false positives. Discriminated abandoned objects and static persons successfully
Maddalena et al. [25]	i-LIDS and Sequence from ViSOR	Static humans in AB Hard caused False detections
Kim et al. [15]	PETS2006	Achieved 91.7% precision and 65.6% recall. Detections are nearly stable in occlusions. False alarms are generated due to background movement
Lin et al. [17]	PETS2006, i-LIDS, own dataset ABODA	Reported 100%precision for PETS and AVSS and 66.67% with ABODA datasets. False alarms are caused by illumination change and crowded scene
Wahyono et al. [19]	PETS2006, i-LIDS, ABODA	Achieved high precision in PETS2006 and AVSS. Scenarios with sudden lighting changes and crowded environment in ABODA are not evaluated
Nam [26]	PETS2006, i-LIDS and PETS2007	Achieved precision of 0.98 and recall of 0.82
Cuevas et al. [20]	PETS2006, i-LIDS, LASIESTA3, ChangeDetection	Most accurate detections, also avoids inaccurate classifications
Ilias et al. [21]	PETS2006, i-LIDS ChangeDetection, ABODA	Edge-based method gives considerably good results in lighting change scenes
Sidyakin et al. [23]	St. Petersburg Metro (own dataset)	Works well except when complex background, large people flow, view angle
Smeureanu and Ionescu [24]	i-LIDS, PETS2006, PETS2007, TCD	Better results than baseline of static object detection and CNN

Future research in the field is expected to come up with a solution that works in scenarios with sudden illumination changes and overcrowding. A better adaptive background model could be incorporated. Also, extending the existing methods to a multi-camera environment can be a problem to work on. Since the integrated details from different viewpoints could reduce the ambiguity and improve performance.

References

1. Role of CCTV cameras: public, privacy and protection. https://www.ifsecglobal.com/
2. PETS 2006 dataset (2006) http://www.cvg.reading.ac.uk/PETS2006
3. Cuevas C, Martnez R, Garca N (2016) Detection of stationary foreground objects: a survey. Comput Vis Image Understand 152:41–57
4. Porikli F, Ivanov Y, Haga T (2008) Robust abandoned object detection using dual foregrounds. EURASIP J Adv Signal Process
5. Porikli F, Tuzel O (2005) Bayesian background modeling for foreground detection. In: Proceedings of the 3rd ACM international workshop on video surveillance & sensor networks (VSSN '05). Singapore, pp 55–58
6. Bhargava M, Chen C-C, Ryoo M, Aggarwal J (2009) Detection of object abandonment using temporal logic. Mach Vis Appl 20(5):271–281
7. Chen C-C, Aggarwal JK (2008) An adaptive background model initialization algorithm with objects moving at different depths. In: IEEE international conference on image processing (ICIP). San Diego
8. Chang JY, Liao HH, Chen LG (2010) Localized detection of abandoned luggage. EURASIP J Adv Signal Process 2010:1–10
9. Tian Y, Feris RS, Liu H, Hampapur A, Sun M-T (2011) Robust detection of abandoned and removed objects in complex surveillance videos. IEEE Trans Syst, Man, Cybern C, Appl Rev 41(5):565–576
10. Stauffer C, Grimson WEL (1999) Adaptive background mixture models for real-time tracking. CVPR99
11. Fan Q, Pankanti S (2011) Modeling of temporarily static objects for robust abandoned object detection in urban surveillance. In: Proceedings of the 8th IEEE international conference on AVSS, pp 36–41
12. Tian Y, Lu M, Hampapur A (2005) Robust and efficient foreground analysis for real-time video surveillance. In: CVPR
13. Chang L, Zhao H, Zhai S, Ma Y, Liu H (2013) Robust abandoned object detection and analysis based on online learning. In: IEEE international confernece on robotics and Biomimetics, pp 940–945
14. Grabner H, Bischof H (2006) On-line boosting and vision. Proc CVPR 1:260–267
15. Kim J, Rivera AR, Ryu B, Ahn K, Chae O (2014) Unattended object detection based on edge-segment distributions. In: IEEE international conference on advanced video signal based surveillance, pp 283–288
16. Kim J, Murshed M, Ramirez Rivera A, Chae O (2013) Background modelling using edge-segment distributions. Int J Adv Rob Syst 10
17. Lin K, Chen S-C, Lin D-T, Hung Y-P (2015) Abandoned object detection via temporal consistency modeling and back-tracking verification for visual surveillance. IEEE Trans Inform Forensic Secur 10(7):1359–1370
18. Stauffer C, Grimson WEL (1999) Adaptive background mixture models for real-time tracking. In: Proceedings of the IEEE computer society conference on CVPR, vol 2, pp 246–252
19. Wahyono, Filonenko A, Jo K-H (2016) Unattended object identification for intelligent surveillance systems using sequence of dual background difference. IEEE Trans Ind Inform 12(6):2247–2255

20. Cuevas C, Martinez R, Berjon D, Garcia N (2017) Detection of stationary foreground objects using multiple nonparametric background-foreground models on a finite state machine. IEEE Trans Image Process 26(3):1127–1142
21. Ilias D, El Mezouar, Taleb N, Elbahri M (2017) An edge-based method for effective abandoned luggage detection in complex surveillance videos. Comput Vis Image Understand 158(C):141–151
22. Gruenwedel S, Van Hese P, Philips W (2011) An edge-based approach for robust foreground detection. In: Blanc-Talon J, Kleihorst R, Philips W, Popescu D, Scheunders P (eds) Advanced concepts for intelligent vision systems. ACIVS. Lecture Notes in Computer Science, vol 6915. Springer, Berlin, Heidelberg
23. Sidyakin SV, Vishnyakov BV, Real-time detection of abandoned bags using CNN. In: Proceedings of SPIE, vol 10334
24. Smeureanu S, Ionescu RT (2018) Real-time deep learning method for abandoned luggage detection in video. CoRR, abs/1803.01160: n. pag
25. Maddalena L, Petrosino A (2013) Stopped object detection by learning foreground model in videos. IEEE Trans Neural Netw Learn Syst 24(5):723–735
26. Nam Y (2016) Real-time abandoned and stolen object detection based on spatio-temporal features in crowded scenes. Multimed Tools Appl 75(12):7003–7028
27. AVSS 2007 dataset (2007) http://www.eecs.qmul.ac.uk/andrea/avss2007_d.html
28. Cotres C, Vapnik V (1995) Support-vector networks. Mach Learn 20:273–297
29. Breiman L (2001) Random forests. Mach Learn 4:5–32
30. Cuevas C, Yez EM, Garca N (2016) Labeled dataset for integral evaluation of moving object detection algorithms: LASIESTA. Comput Vis Image Understand 152:103–117
31. Goyette N, Jodoin P-M, Porikli F, Konrad J, Ishwar P (2012) Changedetection.net: a new change detection benchmark dataset. In: Proceedings of the IEEE conference on computer vision pattern recognition workshops, pp 1–8
32. PETS2007. http://www.cvg.reading.ac.uk/PETS2007/data.html

Malware Detection Using Machine Learning Approach

Naresh Babu Muppalaneni and Ripon Patgiri

Abstract These days, Malware has become a significant and developing warning to safety, identification of malware has become tough. Numerous experiments were conducted on intelligent malware identification by using Machine Learning and Data Mining techniques. However, great outcomes have been produced with these practices, most of them were constructed on shallow learning architectures, which are still a little bit uneasy for Malware Identification issues. Malware regularly changes its recognizable characteristics in order to fool identification using typical signature-based models. Behavior-based Malware identification evaluates not only on the signature of the file, but also based on the activity it wants to plan, that is also before it actually executes that action. The static analysis extracts the features without the execution of executable files. But the dynamic analysis executes the feature with execution. In this paper, we consider malware as an opcode sequence. It is identified by using Machine Learning techniques. N-gram opcode sequence is generated from the opcodes of malware and benign files. Using the N-gram opcode sequence, features of the executable files are identified. Using various Machine Learning methods, we predicted whether the executable is a malware or not.

Keywords Machine learning · Malware detection · N-gram opcode · SVM · Decision tree · Nave Bayes

N. B. Muppalaneni (✉) · R. Patgiri
National Institute of Technology Silchar, 788010 Silchar, India
e-mail: nareshmuppalaneni@gmail.com
URL: http://cs.nits.ac.in/naresh

R. Patgiri
e-mail: ripon@cse.nits.ac.in
URL: http://cs.nits.ac.in/rp

R. Patgiri et al. (eds.), *Proceedings of International Conference on Big Data, Machine Learning and Applications*, Lecture Notes in Networks and Systems 180,
https://doi.org/10.1007/978-981-33-4788-5_18

1 Introduction

Malware is a piece of software which is intended to cause harm to your system or network. Malware is different from normal programs in which most of them have the ability to spread itself in the network, remain undetectable, cause damage to the infected system or network persistence. Malware attacks can be traced back to the time, even before the internet became widespread. There are different types of malware examples: Trojan, virus, Spam, Worm, Spywares, Rootkits, Adware, Ransomware. Since there is a great loss caused by Malware, identification of Malware has become the maximum task. Signature-based method is the most broadly used method for Malware identification. Short strings of bytes that are unique to programs are known as Signatures. This method uses a plain pattern-matching approach to detect harmful code. High accuracy can be obtained with this method. However, the signature is sensitive to moderate changes in harmful code. Signature-based approaches cannot identify the modified or previously unseen Malware. Therefore, one basic important issue faced by the antivirus community is how to identify previously unseen harmful code [19].

Intelligent Malware identification approaches are proposed to address this issue. The intelligent Malware identification approach uses Data mining or Machine Learning methods to identify malware. Machine Learning algorithms are used to study the underlying patterns from a given training set, which have both benign and malicious samples. This underlying pattern makes the difference between malicious and benign samples. Saeed et al. [10] provide an up to date comparative study for most of the malware files, and also it concludes a number of malware detection algorithms. It also gives detailed information about malware infection and improves anti-malware systems. Mathur et al. [6] explain about dynamic, static, and hybrid analysis. The static analysis extracts the features without the execution of executable files. But the dynamic analysis executes the feature with execution. Shabtai et al. [13] has used heuristic-based detection. This approach has detected the existing malware and cannot detect newly unknown malware. In this paper, we consider malware as opcode sequence. It is identified by using Machine Learning techniques. N-gram opcode sequence is generated from the opcodes of malware and benign files. Using the N-gram opcode sequence, features of the executable files are identified. Using various Machine Learning methods, we predicted whether the executable is a malware or not.

The key objectives of the paper is outlined below

- To extract features from executable files.
- To detect malware in executable files.
- To conduct extensive experiments on malware using machine learning models.

This paper is organized as follows. Section 2 surveys the background and related work. Section 3 describes the methodology used to experiment our proposed system. Section 4 demonstrates the results of our experiment. And finally, the paper draws a suitable conclusion in Sect. 5.

Table 1 Malware analysis-feature extraction techniques

Year	Authors	Structure	Feature	
			Body	Header
2015	DLLMiner [7]	Tree	–	DLL
2014	Nissim et al. [9]	N-gram	BYT	BYT
2013	Santos et al. [12]	Sequence	OP	–
2012	Jacob et al. [5]	N-gram	BYT	STC
2011	Nataraj et al. [8]	–	BYT	BYT
2010	Sami et al. [11]	Itemset	–	API
2009	Tabish et al. [15]	N-gram	BYT	BYT
2009	PE-Miner [14]	–	STC	STC
2009	Hu et al.[4]	Graph	FC	–
2009	Griffin et al. [3]	Sequence	BYT	BYT
2008	Ye et al. [18]	Itemset	–	API

FC: Function Call, **BYT**: Byte code
API: Application Programming Interface
OP: Operation code, **STC**: Structural features

2 Background

Since the malware changes its behavior and signature, it is difficult to identify it. To predict that an executable has malicious code, we have to develop a machine learning model. But feature extraction is important. Alazab et al. [2] has presented a method for feature extraction based on the API calls in the executable. The authors analyze the malicious behavior by mapping the API calls in the executable with MSDN library.

Tian et al. [16] has presented a new method for identifying the malware by separating the API calls and their parameters. They store the strings in the global list and compare the strings in each executable with the global list.

Ahmadi et al. [1] has done an extensive survey on various techniques of the feature extraction from executables. He also presented feature fusion, combining features and run the machine learning algorithm (Table 1).

3 Methodology

There are different ways of extracting features from malware executables. Since each executable is of different size, each executable has different number of features. The behavior of malware depends on the code injected in the executable. Opcode(operational codes in assembly language) tells how the program behaves. One of the techniques for extracting the features of executable file is opcode sequences. The features of the executable are a collection of opcode n-gram sequence, based on

Table 2 3-gram opcode sequence of Algorithm 1

LDA,MOV,JMP
MOV,JMP,ADD
JMP,ADD,INC
ADD,INC,JE

which we can predict the the behavioral characteristics of executable. The system consists of three major components: opcode sequence, the feature extraction, Malware identification module. The opcode sequence and the character extraction are considered for analyzing program behavior and extracting their behavioral characteristics. The Malware identification module is considered for predicting the behavior of Malware. In this work, we have generated a 3-gram opcode for the executable file. We are not considering the operands to the opcodes. A 3-gram code generated for the Opcode Sequence: LDA, MOV, JMP, ADD, INC, JE is shown in Table 2.

Algorithm 1 Feature extraction from executable

```
1: procedure OPCODE(exe)
2:     for generating n-gram opcode sequence initialize N value
3:     i = 1
4:     loop
5:     generate the N opcodes from i
6:     i++
7:     endloop
8: end procedure
```

For generating the n-gram opcode sequence, we also consider the call graph, which describes about the control flow of execution of instructions. For generating the n-gram opcode sequence, feature extraction is used by using the Malware and benign executables. We have used the dataset from the UCI Machine Learning Repository. The dataset contains features of 301 Malware executable files, 72 benign executables files, and totally 531 attributes. Machine learning model is built for predicting whether the given executable is benign or malware. For that, the dataset is divided into training and test sets and apply the Machine Learning algorithm like SVM, Naive Bayes, decision tree algorithms.

4 Results

We have taken the pre-defined dataset from [17]. The dataset consists of 531 attributes having 301 malware executables and 72 benign executables. For generating the machine learning model, the malware feature dataset is divided into test set and training set. Since the dataset consists of more records about malware executables,

Table 3 Confusion matrix of SVM

		Predicted	
		Malware	Benign
Actual	Malware	151	1
	Benign	1	34

Table 4 Confusion matrix of Decision Tree

		Predicted	
		Malware	Benign
Actual	Malware	152	1
	Benign	1	32

Table 5 Confusion matrix of Navie Bayes

		Predicted	
		Malware	Benign
Actual	Malware	150	1
	Benign	3	33

we split the dataset equally for training and testing. Training set consists of -50% of malware and 50% of benign and test set consists of -50% of malware and 50% of benign. Various Machine Learning techniques were used to build the model using SVM, Nave Bayes, Decision Tree. The results are as follows (Tables 3, 4, and 5).

5 Conclusion

We have used the Machine Learning technique to identify the malware. The Signature-based method is sensitive to a slight change in malware. The static analysis extracts the features without the execution of executable files. But the dynamic analysis executes the feature with execution. We, proposed a machine Learning algorithm that detects the malware with high accuracy. In this paper, we consider malware as opcode sequence. N-gram opcode sequence is generated from the opcodes of malware and benign files. Using the N-gram opcode sequence, features of the executable files are identified. Using various Machine Learning methods, we predicted whether it's a malware or not. We build a machine learning model using SVM, Nave Bayes, Decision Tree to predict whether the executable is benign or malware. Using SVM model, it is reported that the prediction of malware is of 98.93% accuracy. Using Decision Tree model, it is reported that the prediction of malware is of 98.39% accuracy. Using Nave Bayes model, it is reported taht the prediction of malware is of

98.93% accuracy. Since the dataset contains more malware files than that of benign files it is reported high accuracy towards malware class. In this study, the accuracy obtained from the experiment is only opening results that can pave the way for further tests and research within this area.

References

1. Ahmadi M, Ulyanov D, Semenov S, Trofimov M, Giacinto G (2016) Novel feature extraction, selection and fusion for effective malware family classification. In: Proceedings of the sixth ACM conference on data and application security and privacy. ACM, pp 183–194
2. Alazab M, Venkataraman S, Watters P (2010) Towards understanding malware behaviour by the extraction of api calls. In: 2010 second cybercrime and trustworthy computing workshop. IEEE, pp 52–59
3. Griffin K, Schneider S, Hu X, Chiueh T-C (2009) Automatic generation of string signatures for malware detection. In: International workshop on recent advances in intrusion detection. Springer, pp 101–120
4. Hu X, Chiueh T-C, Shin KG (2009) Large-scale malware indexing using function-call graphs. In: Proceedings of the 16th ACM conference on computer and communications security. ACM, pp 611–620
5. Jacob G, Comparetti PM, Neugschwandtner M, Kruegel C, Vigna G (2012) A static, packer-agnostic filter to detect similar malware samples. In: International conference on detection of intrusions and malware, and vulnerability assessment. Springer, pp 102–122
6. Mathur K, Hiranwal S (2013) A survey on techniques in detection and analyzing malware executables. Int J Adv Res Comput Sci Softw Eng 3(4)
7. Narouei M, Ahmadi M, Giacinto G, Takabi H, Sami A (2015) Dllminer: structural mining for malware detection. Secur Commun Netw 8(18):3311–3322
8. Nataraj L, Karthikeyan S, Jacob G, Manjunath B (2011) Malware images: visualization and automatic classification. In: Proceedings of the 8th international symposium on visualization for cyber security. ACM, p 4
9. Nissim N, Moskovitch R, Rokach L, Elovici Y (2014) Novel active learning methods for enhanced pc malware detection in windows OS. Expert Syst Appl 41(13):5843–5857
10. Saeed IA, Selamat A, Abuagoub AM (2013) A survey on malware and malware detection systems. Int J Comput Appl 67(16)
11. Sami A, Yadegari B, Rahimi H, Peiravian N, Hashemi S, Hamze A (2010) Malware detection based on mining api calls. In: Proceedings of the 2010 ACM symposium on applied computing. ACM, pp 1020–1025
12. Santos I, Brezo F, Ugarte-Pedrero X, Bringas PG (2013) Opcode sequences as representation of executables for data-mining-based unknown malware detection. Inf Sci 231:64–82
13. Shabtai A, Moskovitch R, Feher C, Dolev S, Elovici Y (2012) Detecting unknown malicious code by applying classification techniques on opcode patterns. Secur Inform 1(1):1
14. Shafiq MZ, Tabish SM, Mirza F, Farooq M (2009) Pe-miner: mining structural information to detect malicious executables in realtime. In: International workshop on recent advances in intrusion detection. Springer, pp 121–141
15. Tabish SM, Shafiq MZ, Farooq M (2009) Malware detection using statistical analysis of byte-level file content. In: Proceedings of the ACM SIGKDD workshop on cybersecurity and intelligence informatics. ACM, pp 23–31

16. Tian R, Islam R, Batten L, Versteeg S (2010) Differentiating malware from cleanware using behavioural analysis. In: 2010 5th international conference on malicious and unwanted software. IEEE, pp 23–30
17. UCI. Detect malacious executable(antivirus) data set. retrieved on 15/04/2019 from https://archive.ics.uci.edu/ml/datasets/Detect+Malacious+Executable(AntiVirus)
18. Ye Y, Wang D, Li T, Ye D, Jiang Q (2008) An intelligent pe-malware detection system based on association mining. J Comput Virol 4(4):323–334
19. Yuxin D, Siyi Z (2019) Malware detection based on deep learning algorithm. Neural Comput Appl 31(2):461–472

Comparative Study of Faster Region-Based Convolutional Neural Networks with Inception V2 and Single Shot Detector with Inception V2 on Their Signature Detection Capabilities

Ashutosh Bajpai, Sai Kiran Wupadrasta, and Balasubramanian

Abstract Signature detection and verification in offline handwritten signature printed document is a problem when solved helps to target forgery and reduce man hours. With challenges like stray marks, several methods were proposed and eventually with the rise of vision recognition methods and Deep Learning, application program interfaces have been developed to detect signatures. Using TensorFlow APIs, images that are converted to radio frequency (RF) format are processed to detect signatures. This paper intends to evaluate two different object detection algorithms, faster region-based convolutional neural networks and single shot detector with inception V2, on signature dataset. Various standard metrics such as train and test time, loss rate and precision at area under curve etc. are used to compare their performance and fitment for the problem statement.

Keywords Signature detection · Object detection · Faster region-based convolutional neural networks (faster RCNN) · Single shot detector (SSD) · Inception V2

1 Introduction

Security applications are widely using biometrics technologies around the world for authentications. A person is identified using physiological or behavioral traits he/she poses. In the first case, the recognition is based on biological traits, such as fingerprint, face, Iris, etc. Voice and handwritten signatures fall under the category of behavioral traits [1]. Handwritten signature has become the one of the major biometric traits to recognize a person in legal, financial and administrative domain. Signatures are widely used in the process because of their noninvasiveness and familiarity with them in day-to-day life [2]. Even in the times when electronic data are in drastic increase in banking applications, an offline handwritten signature printed document is still

A. Bajpai (✉) · S. K. Wupadrasta · Balasubramanian
Indian Institute of Technology Hyderabad, Hyderabad, India
e-mail: ashutoshbajpai.ma.gre@gmail.com

R. Patgiri et al. (eds.), *Proceedings of International Conference on Big Data, Machine Learning and Applications*, Lecture Notes in Networks and Systems 180,
https://doi.org/10.1007/978-981-33-4788-5_19

Fig. 1 Example signature document

considered to be most reliable way of user commitment, agreement and identification. An illustrative example is shown in Fig. 1. A manual inspection is carried out by the administrative officers by scanning a signed document, which is to be processed. The errors and false interpretations are associated with manual inspection. Moreover, the additional workload either leads to the overall increased cost in terms of human resource or working hours. Therefore, it is more efficient to use intelligent software techniques for automated analysis of documents to perform signature related tasks [3]. Exploration of large set of heterogeneous documents images can be easily enabled using index provided by the signature in multiple domains such as forensic and banking where uniqueness and evidence are key aspects. Individually authorized documents retrieval can be effectively carried out by locating a specific signature, given an abundance of documents [4]. The study of offline signature detection and verification has been broadly researched in the last decades but remains an open research problem in the current times too. The objective of signature detection and verification systems is to detect if a given signature is authentic (generated by the claimed person), or forged (generated by a fraudster) by appropriately locating it in a document followed by verification. Although, there have been significant advancements achieved on signature verification systems, signature detection on various types of documents still pose a serious challenge due to large contribution by its vast inherent unstructured-ness.

1.1 Challenges

There are numerous challenges in detecting offline signature in scanned documents. Low resolution is the first problem in these types of documents, which makes them difficult to enhance. Second, the background texts of each document are different and usually not known beforehand. Third, urgency of application poses a challenge by making documents subject to restricted processing time. Most importantly, the document images often contain noisy lines or dots and other handwritten information that resemble or overlap with signatures.

1.2 Faster R-CNN

Faster R-CNN can be identified as canonical model for object detection based on deep learning. A standard process flow of faster RCNN is presented in Fig. 2. Three simple steps can effectively describe R-CNN, or Region-based Convolutional Neural Network. First, an input image is scanned and approximately 2000 region proposals are made by selective search method for possible objects. Second, these region proposals are fed into the CNN (Convolutional Neural Network). At last, the outcome of each CNN is fed into two different models—(a) a classifier is used to classify the region proposal and (b) a linear regressor is used to tighten the bounding box of the object, if such an object exists. Slow selective search algorithm has been replaced with a fast-neural net in Faster R-CNN, which is the improvement over R-CNN. Specifically, faster RCNN introduced the Region Proposal Network (RPN).

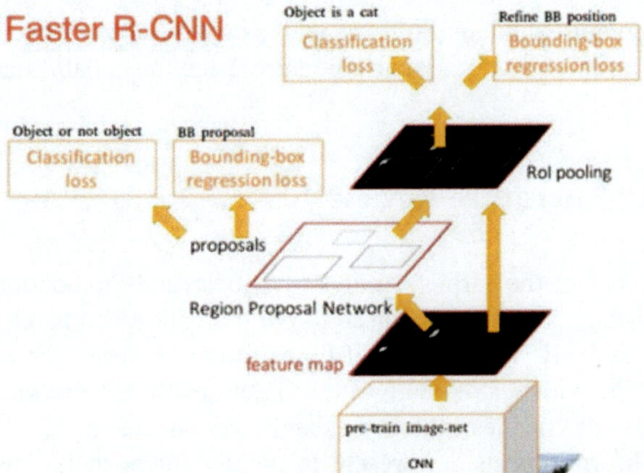

Fig. 2 Faster RCNN architecture

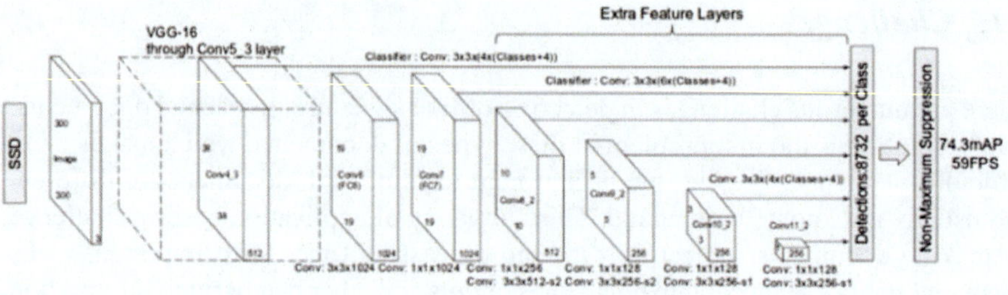

Fig. 3 SSD architecture from original paper

1.3 Single Shot Detector (SSD)

SSD stands for Single Shot Detector. A standard process flow of SSD is demonstrated in Fig. 3. It outperforms Faster RCNN when compared with execution speed. Unlike RCNN, region proposals and classification are done in a "single shot" simultaneously predicting the bounding box and the class as it processes the image in case of SSD.

1.4 Objective

The objective of this study is to compare two new generation pretrained networks faster R-CNN with inception V2 and SSD with Inception V2, respectively, on their signature detection capabilities on signature-based documents. Signature detection and verification has been in research for quite a long time. However, recent advancements in vision recognition algorithms have not been significantly explored on signature domain. Study as well as comparison of two benchmark pretrained vision recognition networks on signature detection capabilities have not been explored yet.

2 Literature Review

In one of the earliest studies by Djeziri and Plamondon [5], the imitation of human visual perception was used to build an approach tackle signature extraction process intuitively. The concept of Filiformity was described as a criterion for the curvature characteristics of handwritten signatures in their work. This concept was successful in bank cheques, which are usually very neat and clean. However, it failed when other filiform objects are present in the document including signatures. Sliding window concept was introduced by Madasu and Lovel [6] to crop the image segment by estimating the area where the signature lies. Then local entropy derived from the pixel-based density of the region was used to decide if it is a signature or not. High-density regions were reported as signatures incorrectly due to the disregard for the noise. Geometric features such as area, circularity, aspect ratio, size and position were also used to analyze segmented regions by Madasu et al. [7] and Chalechale et al. [8]. Pairwise similarity metrics such as Manhattan distance were used to compare the similarity of two signatures as well. Jayavdean et al. [9] also tried an innovative method of variance analysis in a grid placed on the putative signature position in gray-scale images mostly cheques. A three-step method to extract signatures was introduced by Mandal et al. [12]. First, word-level feature extraction used to locate the signature segment. Second, dehyphenation is performed between signature strokes and background text which overlaps. In final step, signature strokes get classified using Skelton analysis. Gradient based features were also used as input to learning-based classifiers. Zhu et al. introduced the concept of multi scale saliency feature to describe signature characteristics [10] and, then it was subsequently used for signature verification process also [11]. Esteban et al. [14] approached the problem differently as they assumed that the signature detection is followed by signature verification therefore, a known dataset of signatures was used to build an evidence accumulation strategy. This process was highly accurate but, given assumption is not always true since the bank application forms are often filled by new or perspective customers and this process is followed by archiving not by verification. Speeded Up Robust Features (SURF) was also used by Ahmed et al. [13] to classify image block segments.

Some key survey papers also tried to summarize the advancements in this field over the last few decades, in the late 80 s [15], 90 s [16] and 2000s [17]. Acquisition devices also were in focus specifically in case of online signatures and its representation. Here are some literature reviews, which consolidated the recent advancement: Impedovo et al. [18] provide an update over the authors' previous review [17]; Shah et al. [19] present a critical evaluation of 15 signature verification systems proposed in the literature, classifying each work according to the feature extraction methods, classifiers and overall strengths and limitations of the systems. Although, there has been significant work done on several methods for signature detection, the latest developments involving the usage of Deep Learning methods or object detection algorithms has not been explored much for handwritten signatures. Some of those methods have demonstrated superior results for object detection in multiple benchmarks and are reviewed in this paper.

Figure 4 shows the benchmark comparison report of both algorithms done at Google Research by Huang et al. [20]. It can be inferred from this paper that faster RCNN is the better choice given no constraints on speed while SSD is better for real time speed. Faster RCNN is relatively better for small objects detection.

3 Methodology

Dataset preparation is the first step in the methodology. In this study, the dataset is randomly split into 80–20 ratio for Training and Testing. Tensor flow object detection APIs are widely used in object detection related research. This paper evaluates

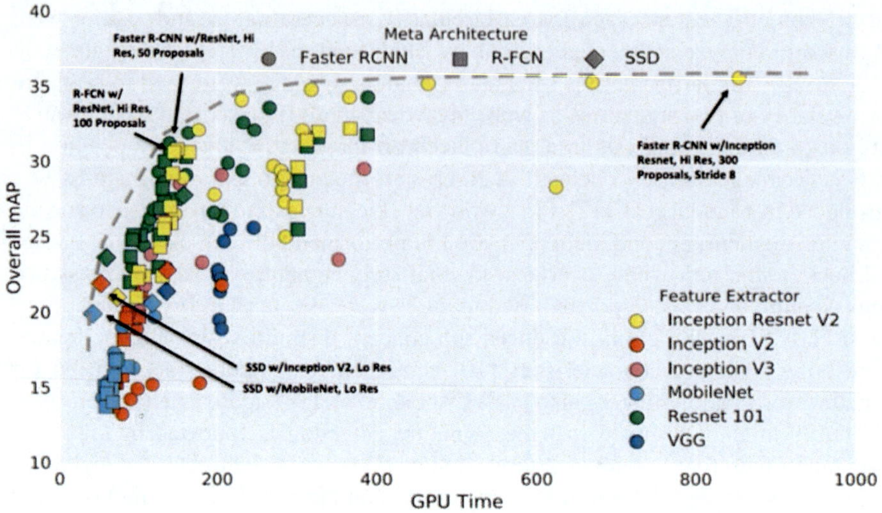

Fig. 4 Comparative analysis of faster RCNN and SSD | google research

two algorithms named faster R-CNN with inception v2 and SSD with inception v2 pretrained on coco dataset. The input format of the training data is TF-records. TF-records are specific standard format used to train any algorithm in Tensor Flow object detection API. TF records are generated from the CSV annotated data using an automated script. To generate the annotated data, we use LabelImg, a third-party open source UI-based tool to label multiple objects in image. It generates XML as an output for annotated data, which can subsequently be transformed into CSV format.

There is a configuration file containing several parameters such as number of epochs, learning rate, input image size etc. that can be tuned. A model checkpoint is exported after the training completion followed by the inference graph frozen process to freeze the graph and export it for prediction process. These exported inference graphs get imported in prediction process to detect objects in test images. An architectural process flow is attached here. The workflow architecture is given in Fig. 5.

The key performance metrics that will be used to compare the two algorithms are

- Comparative results on training
  - Loss function
  - Learning rate function
  - Execution overhead
- Comparative results on testing
  - Precision versus Recall curve using IOU(Intersection over Union) threshold
  - mAP (mean average precision)
  - Highest precision value

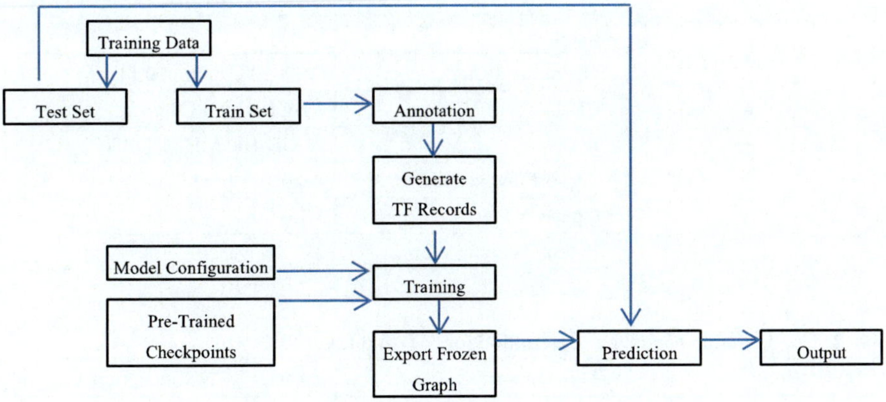

Fig. 5 Workflow architecture

- Execution overhead.

4 Results and Analysis

The comparative study of SSD with inception V2 and Faster RCNN presented here is based on base comparison on fixed configuration and platform setting. Intended outcome of the analysis is to find the better suitability and fitment of one of the two object detection strategies on signature detection capability.

4.1 Dataset

Dataset, which is pilot in nature, comprises of 50 signature document images (~2000*3000 pixels) generated randomly by pasting max 3 signatures (from the set of 22 unique signatures) on documents at random places. Train and test set split rate kept at 80–20. There are couples of images, which do not follow the training set distribution, have been added to test set to evaluate the generality of algorithms.

4.2 Platform and Environment Specification

Platform and environment specification are given in Table 1. GPU was not used in the training process as the dataset was limited number of examples and it was sufficient to use standalone system.

Table 1 Platform description

Parameters	Values
Operating system	OS X El Captain 10.11.6
CPU	2.2 GHz Intel Core i7
RAM	16 GB 100 MHz DDR3
TensorFlow version	1.8.0
Open CV	3.1.0
Python version	2.7.10

Table 2 Configuration parameters description for SSD and faster RCNN

Parameters	SSD inception V2	Faster RCNN inception V2
Pre-trained checkpoint	ssd_inception_v2_coco_2018_01_28	faster_rcnn_inception_v2_coco_2018_01_28
Input dim	512*512	Max 512
Batch size	1	1
Initial learning rate	0.004	0.004
Number of epochs	15 k	15 k
Max detection per class	3	3
Max detection total	3	3
Momentum	0.9	0.9
Score convertor	Sigmoid	Softmax

4.3 Configuration Parameters

Configuration parameters are given in Table 2. Pretrained checkpoints of COCO dataset were available for both the algorithms. These checkpoints have been used to initialize the weights for the training. Batch size was kept at 1 because of the constraint posed by faster RCNN as it keeps the aspect ratio intact while resizing different size input images. The process was run for 15 k epochs, which took roughly 8 h each for both algorithms. As the signatures are of minimal size compared with the whole document, it was required to use sufficient input dimension size which can generate sufficient features. Total number of maximum signatures on a document is kept at 3. As in many cases, a document is approved till level 3.

4.4 Training Results

Training results have been subdivided into four sections; Overall comparison detail, learning rate comparison, loss comparison and histogram of weights and bias for box predictor and classification comparison.

Table 3 Training results for SSD and faster RCNN

Parameters	SSD inception V2	Faster RCNN inception V2
Execution time	436 m 10.682 s	438 m 30.545 s
Average loss (first 40 epoch)	7.365	0.5005
Average loss (last 40 epoch)	3.16	0.0058
Percentage decrease in average loss after 15 k epochs	57%	98%

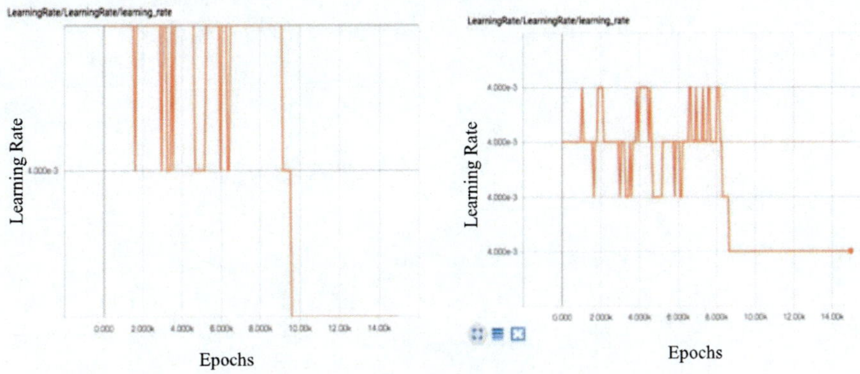

Fig. 6 Learning rate change comparison of SSD and Faster RCNN

Faster RCNN performs better in terms of training outcomes compared to SSD. The percentage loss decrease after 15 k epochs in Faster RCNN was at 98% compared with 57% in SSD. Figure 7 demonstrates the loss value change over epochs. It can be inferred from the percentage loss decrease than the Faster RCNN learning was substantially higher than the SSD. The initial and final loss value is also significantly less in Faster RCNN, which reflects its better fitment for the problem statement. However, the execution time for training process in both algorithms stands at par with each other. These inferences are from Table 3. There was no significant change in learning rate from the initial one as seen from Fig. 6. The drop rate of total loss in faster RCNN was much higher than the SSD. SSD has not converged after 15 k epochs depicting significant loss. The change in values of bias and weights for both box and class predictor in histogram format is presented in Fig. 8 for SSD and Faster RCCN, respectively.

4.5 Testing Results

Trained graphs are generated after the training process for both the algorithms. The unseen set of 20% data was utilized to test the models. A sample output image is

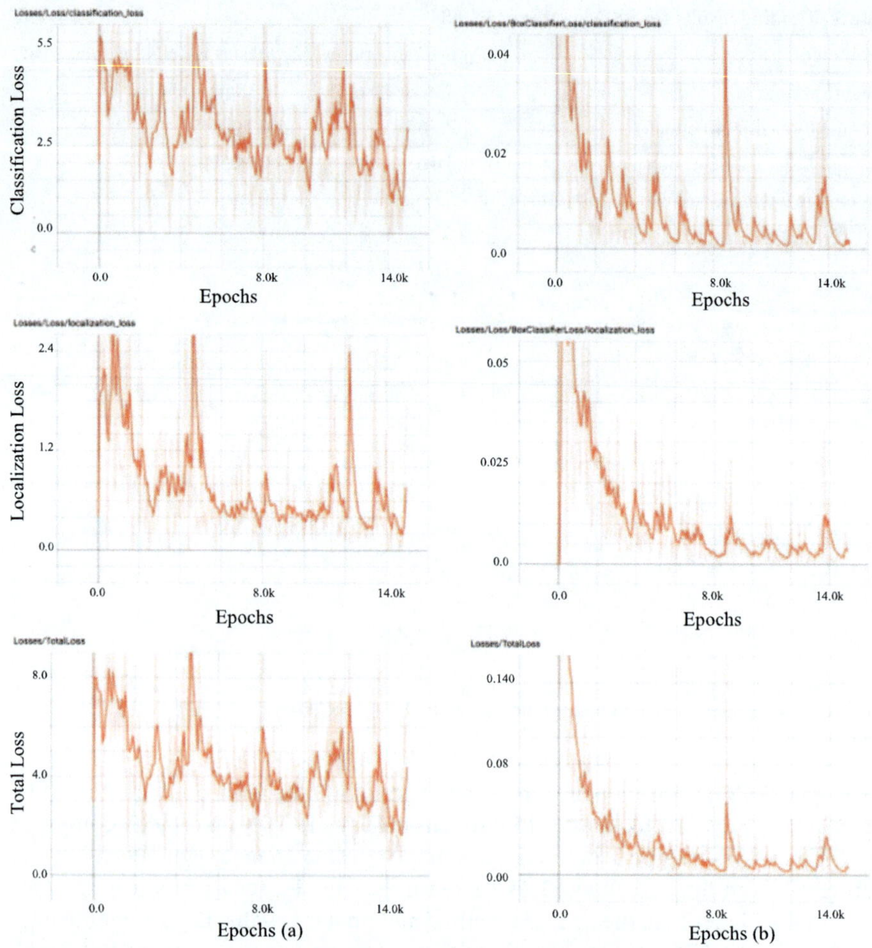

Fig. 7 Loss comparison during training for SSD (**a**) and Faster RCNN (**b**)

shown in Fig. 9. Object detection comprises two core tasks—finding the bounding box and classifying the object. Mean average precision (mAP) has become the base metric to evaluate the performance of object detection algorithms. It uses the IOU (intersection over union) to include the evaluation component for bounding box also. The mean Average Precision or mAP score is calculated by taking the mean AP over all classes and/or over all IoU thresholds (Table 4).

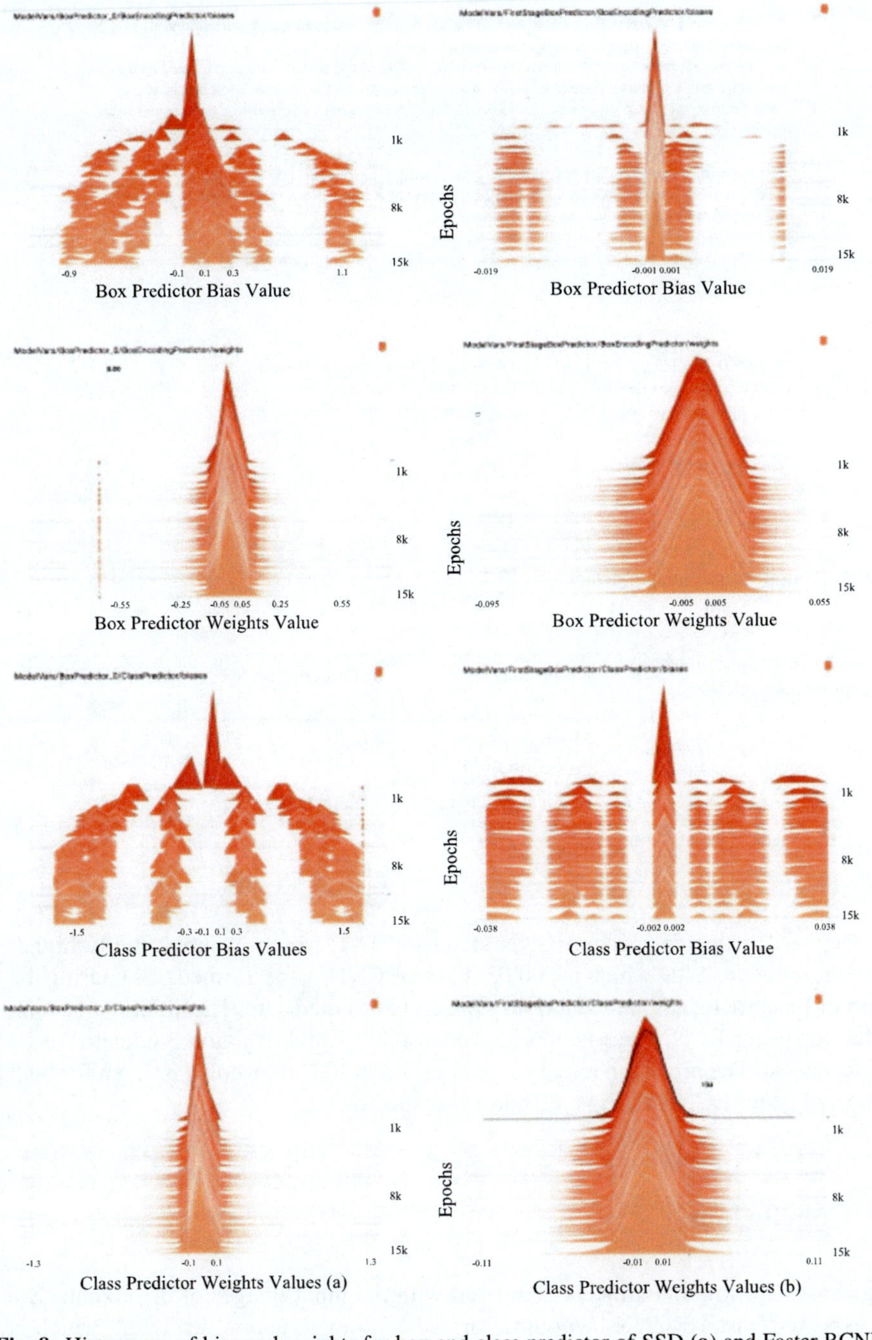

Fig. 8 Histograms of bias and weights for box and class predictor of SSD (**a**) and Faster RCNN (**b**) during training

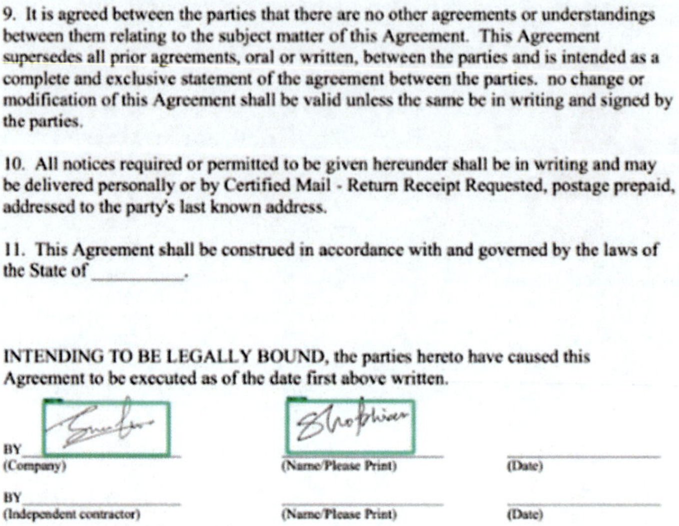
9. It is agreed between the parties that there are no other agreements or understandings between them relating to the subject matter of this Agreement. This Agreement supersedes all prior agreements, oral or written, between the parties and is intended as a complete and exclusive statement of the agreement between the parties. no change or modification of this Agreement shall be valid unless the same be in writing and signed by the parties.

10. All notices required or permitted to be given hereunder shall be in writing and may be delivered personally or by Certified Mail - Return Receipt Requested, postage prepaid, addressed to the party's last known address.

11. This Agreement shall be construed in accordance with and governed by the laws of the State of ________.

INTENDING TO BE LEGALLY BOUND, the parties hereto have caused this Agreement to be executed as of the date first above written.

BY________ ________ ________
(Company) (Name/Please Print) (Date)

BY________ ________ ________
(Independent contractor) (Name/Please Print) (Date)

Fig. 9 Sample output image, detected

Table 4 Testing results for SSD and faster RCNN

Parameters	SSD inception V2	Faster RCNN inception V2
Mean average precision mAP	0.0309	0.47
Highest precision value @ IOU Threshold	0.004@0.5	0.873@0.5

The mAP value for faster RCNN stands at 0.47, which is substantially higher than the value for SSD which is at 0.03. Faster RCNN outperformed SSD during the training phase in loss value reduction, which is to be noted here. The highest precision value for faster RCNN was achieved around 0.87, which is a good indicator of its performance. The precision recall curve at different IOU thresholds is given for both SSD and faster RCNN in Figs. 10 and 11, respectively.

5 Conclusions

Signatures are low size objects compared with the input images of documents. So, as expected, faster RCNN has outperformed SSD in signature detection, which is an observation in line with the inferences from the benchmark comparison report of several algorithms on standard images detection done by Jonathan Huang et al. [20]. The speed comparison is also in line with the same where SSD beats RCNN,

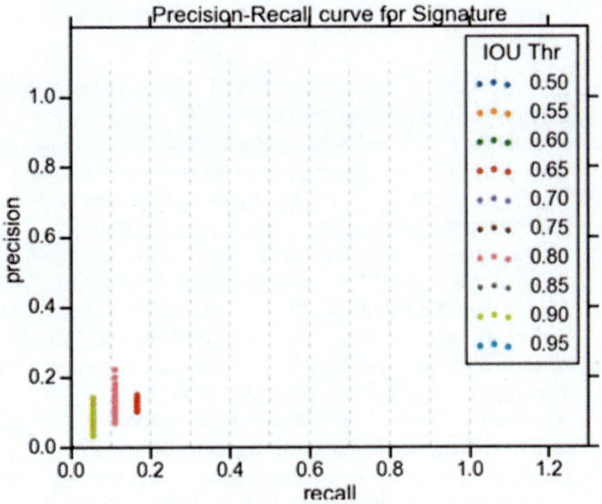

Fig. 10 Precision-Recall curve for SSD at different IOU threshold

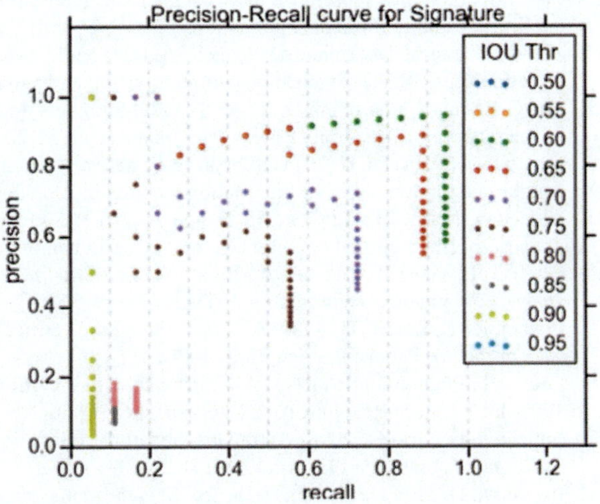

Fig. 11 Precision-Recall curve for faster RCNN at different IOU threshold

as expected from the very architecture of SSD. A full-scale commercial application is feasible using object detection algorithms, specifically faster RCNN for signature detection. Other architectures, like Inception V3 and Inception Resnet v2, within faster RCNN can be explored further to extend this comparative study on signature detection.

References

1. Jain AK, Ross A, Prabhakar S (2004) An introduction to biometric recognition. IEEE Trans Circuits Syst Video Technol 14(1):4–20
2. Plamondon R, Srihari SN (2000) Online and off- line handwriting recognition: a comprehensive survey. IEEE Trans Pattern Anal Mach Intell 22(1):63–84
3. Cüceloğlu İ et al, Detecting handwritten signatures in scanned documents. In: 19th computer vision winter workshop, At Křtiny, Czech Republic
4. Zhu G, Zheng D, Jaeger S (2007) Multi- scale structural saliency for signature detection. In: Proceedings of the IEEE conference on computer vision and pattern recognition, pp 1–8
5. Djeziri FN, Plamondon R (1998) Extraction of signatures from check background based on a filiformity criterion. IEEE Trans Image Process 7(10):1425–1438
6. Madasu VK, Lovel BC (2005) Automatic segmentation and recognition of bank cheque fields. Digit Image Comput Tech Appl 80(1):33–40
7. Madasu VK, Yousuf MHM, Hanmandlu M, Kubik K (2003) Automatic extraction of signatures from bank cheques and other documents. In: DICTA'03
8. Chalechale A, Naghdy G, Prematatne P, Mertins A (2004) Document image analysis and verification using cursive signature. In: IEEE international conference on multimedia and expo
9. Jayadevan R et al (2007) Variance based extraction and hidden Markov model based verification of signatures present on bank cheques. In: International conference on computational intelligence and multimedia applications
10. Zhu G, Zheng Y, Doermann D, Jeager S (2007) Multiscale structural saliency for signature detection. In: IEEE conference on computer vision and pattern recognition
11. Zhu G, Zheng Y, Doermann D, Jeager S (2009) Signature detection and matching for document image retrieval. IEEE Trans Pattern Anal Mach Intell 31:2015–2031
12. Mandal R, Roy PP, Pal U (2011) Signature segmentation from machine printed documents using conditional random field. In: International conference on document analysis and recognition
13. Ahmed S, Malik MI, Liwicki M, Denge A (2012) Signature segmentation from document images. In: International conference on frontiers in handwriting recognition
14. Esteban JL, Velez JF, Sánchez A (2012) Sánchez. Off-line handwritten signature detection by analysis of evidence accumulation. IJDAR 15:359–368
15. Plamondon R, Lorette G (1989) Automatic signature verification and writer identification—the state of the art. Pattern Recogn 22(2):107–131
16. Leclerc F, Plamondon R (1994) Automatic signature verification: the state of the art—1989–1993. Int J Pattern Recognit Artif Intell 8(03):643–660
17. Impedovo D, Pirlo G (2008) Automatic signature verification: the state of the art. IEEE Trans Syst, Man, Cybern, Part C (Appl Rev) 38(5):609–635
18. Impedovo D, Pirlo G, Plamondon R (2012) Handwritten signature verification: New advancements and open issues. In: International conference on frontiers in handwriting recognition. IEEE
19. Shah AS, Khan MNA, Shah A (2015) An appraisal of off-line signature verification techniques. Int J Modern Educ Comput Sci 4:67–75
20. Huang J et al (2017) Speed/accuracy trade-offs for modern convolutional object detectors. arXiv: 1611.10012v3 [cs.CV]

Association Rules Extraction from Date's Product Dataset Using the Apriori Algorithm

Jorge Diaz, David Ovallos-Gazabon, and Carlos Vargas Mercado

Abstract When working with large data collections to obtain, in a short time, the most relevant information they contain, it is possible to perform pattern extraction by means of Data Mining. Among the most used patterns are Association Rules, which measure the co-occurrence of items in large collections of transactions. After many trials, one type of transaction has been found that can be treated more efficiently than the one that has been used up to now. In this study, a new method was applied to this type of transactions, which allowed to obtain in the first tests much faster execution times and more information than the one obtained with the classic Association Rules Mining Algorithms. This will allow to improve the response times of a web recommendation system to provide answers to the users in real time.

Keywords Association Rules Mining (ARM) · Big Data · Recommendation Systems (RS) · Apriori algorithm

1 Introduction

Since the beginning, Association Rules Mining (ARM) [1] has protected itself from the dilemma of the rare item by defining a minimum support that prevents the study of the less supported items (rare items) to avoid memory overflow problems and to be able to offer results in a short time. In the evolution of ARM, there are multiple contributions that alleviate this problem, but always at the cost of avoiding the study of some existing relationships between the data under study.

J. Diaz (✉)
Universidad de la Costa (CUC), Barranquilla, Colombia
e-mail: jdiaz5@cuc.edu.co

D. Ovallos-Gazabon
Universidad Simón Bolívar, Barranquilla, Colombia
e-mail: david.ovallos@unisimonbolivar.edu.co

C. V. Mercado
Corporación Universitaria Latinoamericana, Barranquilla, Colombia
e-mail: cvargas@ul.edu.co

R. Patgiri et al. (eds.), *Proceedings of International Conference on Big Data, Machine Learning and Applications*, Lecture Notes in Networks and Systems 180,
https://doi.org/10.1007/978-981-33-4788-5_20

In the first works on ARM, the aim was to avoid the search for relationships between rare items [1–3]. The use of a relative support for each item set based on the confidence of the rules [4] that generate its items, which allows to save information of the item sets with less support than the minimum support that exceeds their relative support [5], guaranteeing that the rules they generate are of quality. It is proposed in [6] for an improvement of the algorithm proposed in [4] to provide it with scalability. In [7], they work with the same ideas of multiple support, but they reduce the number of itemsets to be saved according to their lift, which measures the improvement produced by the presence of an item in the support of the rest of the items it contains. In [8], they use multiple supports based on the joint distribution of the items and not only on their individual distribution, as proposed in [4]. All these contributions reduce the number of item sets stored in memory, so that other less frequent item sets can be stored, but with better information about the population under study.

When rare items are worked on with large amounts of data, the techniques are usually chess and mushroom [9–11].

All researchers who have worked with these collections have treated their transactions in the same way as the classic "shopping basket" transaction. In this research, it was found that a different treatment can be given without losing the advantages of MRA or applying new algorithms.

Section 2 provides a formal definition for this type of transaction. Section 3 shows a method that solves the dilemma of the rare item in this type of transactions using the classic algorithms of ARM. Section 4 presents the experimental results obtained with Chess and Mushroom, and Sect. 5 presents the conclusions of this study.

2 Type of Transactions

A transaction is a series of elements or characteristics (which will be called items) that are observed together under certain circumstances. This simple definition allows to model many data collections through transactions, from the classic "shopping basket" to the study of a web browsing session, as if it was a simple set of web pages visited together, or the observation of the characteristics of individuals with similar diseases or the detection of credit card fraud.

If it is not casual to form a transaction by grouping some items and not others, and a large set of transactions is observed, it is not difficult to find in the sets of items that repeat, sets that are called frequent sets or patterns. For example, in the shopping cart, these patterns may suggest that "the customer who buys product A usually buys product B together." In the analysis of navigation on a website, there is a rule that suggests that "the user who visits page A usually goes to page B at the same time" or even that "page A must be on the same topic as page B," since users often visit them together [12, 13].

It is important to note that ARM often works without prior knowledge of the population under study, leaving the data to display that information. All ARM algorithms seek to extract the maximum knowledge from a large number of transactions in the shortest possible time. By applying them, adjustments can be made for each specific study, usually based on previous studies done on the same population. However, when considering them, care must be taken to ensure that they are applicable to all the populations under study.

To better understand the mushroom data collection, in its design, 22 different attributes were considered (color, texture...) from a total of 8124 different mushrooms, each one classified as "poisonous" or "edible." In order to give these data, a transaction form and to be able to study them by means of ARM, a unique code is assigned to each of the attribute-value pairs. Thus, codes 1 and 2 correspond to the classification made. Codes 3, 4, 5, 6, 7 and 8 indicate the color. Each mushroom is observed and its classification, color, etc. is recorded by means of a transaction, so that all transactions are composed of 23 items (it's class plus the value of the 22 attributes) [14, 15].

Another characteristic of this type of transaction is the high density of its correlation matrix. In practice, the difficulty of working with transactions of this type is due to the large number of relationships in many data collections used. Each attribute value is related to almost all the values of the other attributes, a fact that does not occur in the classic shopping basket. This is what causes the lack of resources and the excess of time spent when trying to thoroughly analyze these data collections with classic ARM algorithms. If a transaction of this type did not have a dense correlation matrix and had a small number of items, its analysis would not present any problem [16].

In [12], they convert the browsing sessions of users of an e-commerce site into Type II transactions and use the association rules obtained to suggest improvements in the design of the site. The attributes studied in this analysis collect information generated by portal users: browser, type of visitor, keywords, source, etc. All the attributes were divided into a small set of representative values and, in order to form the transactions, the specific value of each attribute was noted during a visit to the website.

Other e-commerce sites can model their data through Type II transactions, so that they can benefit from the advantages proposed in this study. For example, a furniture sales portal could be used in the design of a room. If the user selects a certain model of bed, other furniture could be combined with that model of bed for other users to improve the recommendations of the site to the user. If the analysis is done using sufficiently fast algorithms, recommendations could be made in real time, as the user makes his selections [17].

3 Efficient Analysis of Type II Transactions

Type II transactions can present problems in the analysis when there is a relationship between most of the items. However, before proceeding to their analysis, it is important to note that the different items that form an attribute are, by definition, exclusive. This feature has not been used in any of the previous studies on MRA that were reviewed, so its use is proposed so that the information contained in the data is not lost, and it will also allow to address problems that cannot be solved with the other proposals due to lack of memory resources or because they take much longer than the average user would accept in a Recommendation System [17, 18].

Once these values are removed from the repository under study, any of the algorithms proposed for ARM can be applied, both the Apriori [19] type algorithms with minimum unitary support and those using multiple minimum supports are described in the first section of this paper.

By removing all the proposed items from the Type II transaction repository, the removed information can be retrieved after the analysis, and what the method does is simply to reduce its dimensions, causing a faster and, in many cases, deeper analysis [20].

Suppose there is a population with three attributes under study whose values are coded with {1,2} for the first attribute, {3,4,5} for the second and {6,7} for the third. Assuming that a transaction repository (D) presents maximum variability, in the classic Apriori algorithm the tree L shown in Fig. 1 will be generated to collect all the co-occurrences found in the repository.

Even though it works with millions of transactions in D, the L tree in Fig. 1 keeps all the information required to obtain all the association rules of D.

If the proposal is considered and items 1, 3 and 6 are removed from the repository, the L tree will be reduced to the one shown in Fig. 2.

The number of nodes is reduced from 40 to only 11, which allows to obtain all the existing relations between any pair of items in the reduced repository, in much less time and using much less memory space.

To obtain information about the items that are not present in the L tree obtained, it is enough to refer to the design used in the codification of the values of each attribute under study:

- The support of item 1 is obtained by subtracting the support of item 2 from N. n1 = N – n2
- n3 = N – n4 – n5
- n23 = n2 – n24 – n25
- n13 = n3 – n23
- …

The support of any node in the entire L-tree is obtained by a couple of targeted searches in the reduced tree and a difference, so it hardly consumes any time or memory resources.

3.1 Refinement

If the time spent on the analysis is to be further reduced, the reduction of D-items can be forced to a maximum without changing the obtained results. In the previous example, the items 1, 3 and 6 have been eliminated at random. If the first reading

Fig. 1 L Tree

1 (n_1)	-	2 (n_{12})		
	-	3 (n_{13})	-	6 (n_{136})
			-	7 (n_{137})
	-	4 (n_{14})	-	6 (n_{146})
			-	7 (n_{147})
	-	5 (n_{15})	-	6 (n_{156})
			-	7 (n_{157})
	-	6 (n_{16})		
	-	7 (n_{17})		
2 (n_2)	-	3 (n_{23})	-	6 (n_{236})
			-	7 (n_{237})
	-	4 (n_{24})	-	6 (n_{246})
			-	7 (n_{247})
	-	5 (n_{25})	-	6 (n_{256})
			-	7 (n_{257})
	-	6 (n_{26})		
	-	7 (n_{27})		
3 (n_3)	-	4 (n_{34})		
	-	5 (n_{35})		
	-	6 (n_{36})		
	-	7 (n_{37})		
4 (n_4)	-	5 (n_{45})		
	-	6 (n_{46})		
	-	7 (n_{47})		
5 (n_5)	-	6 (n_{56})		
	-	7 (n_{57})		
6 (n_6)	-	7 (n_{67})		
7 (n_7)				

Fig. 2 Reduced L-tree

2 (n_2)	-	4 (n_{24})	-	7 (n_{247})
	-	5 (n_{25})	-	7 (n_{257})
	-	7 (n_{27})		
4 (n_4)	-	7 (n_{47})		
5 (n_5)	-	7 (n_{57})		
7 (n_7)				

of D is used to check which item is more frequent in each attribute, it is possible to select that item to eliminate it from the study, so that the size of D is further reduced without losing any useful information.

This reduction of D will mean that in each transaction there will be fewer items to study and therefore far fewer item sets to go through to count its support and fewer searches in the L tree to update it. These are the most time-consuming processes in ARM algorithms so, without losing information, there will be a gain in response time when performing the analysis.

This reduction may also lead to a smaller size of the L tree, since the items eliminated are the most frequent and, therefore, those that tend to have the most relationships with other items in the study.

3.2 *Unrelated Items*

In the classical analysis for obtaining association rules, it is frequent to use a minimum support higher than 0% to complete it. This causes that that if no relation is found between two particular items, it can be caused by two different issues:

- The number of times they are related in B is lower that the established minimal support.
- They are not really related in any transaction of D.

4 Experimentation

This type of transactions, specifically chess and mushroom, has been used in previous experiments available in the UCI Machine Learning Repository FIMI, but it has always been necessary to forego complete data analysis due to the high co-occurrence of its items and the large number of long itemsets they produce. In other consulted studies that use these data, they have also renounced to the complete analysis of the repository using minimum support higher than 0%.

The characteristics of the experiments can be seen in Table 1. When executing classic Apriori [2], the minimum support was 0. This value was increased until the algorithm was completely executed, without overflowing the memory. Then, the proposal is used, eliminating in case (1) a value for each randomly selected attribute, and in case (2) eliminating the most present value in D for each attribute. In both cases, it was possible to work with a minimum support of 0%, obtaining information about 100% of the items under study.

The important thing in both cases is that it was possible to carry out a complete analysis without renouncing the information that the data have on all the items under study and in a reasonable time.

Table 1 Comparison between the classic algorithm and this proposal

	Classic Apriori			Type II (1)	Type II (2)
	Minimum support (%)	Rare items (%)	Time (sg)		
Chess	40	34.4	8524	201	124
Mushroom	2	25%	3521	189	154

In addition to finding all the relationships involving rare items, it is possible to detect which items are not related, something that was not possible with the classic algorithms due to the use of minimum not null support.

5 Conclusion

Experiments with the transactions defined as type II show that it is possible to obtain a great increase in efficiency of the ARM algorithms when analyzing this type of transactions. The proposed method implies a reduction of a large amount of the data to be processed without any loss of the information they contain, allowing to apply any of the existing ARM algorithms to the new data. Even in many experiments, it is possible to discover what some authors call negative association rules, i.e., the absence of a relationship between certain items in the repository under study, without having to modify the search criteria of the algorithms.

References

1. Grupo González Giménez Desde 1952. https://goo.gl/MY3oVv. Accedido 05 Jul 2018
2. Silva J, Cubillos J, Villa JV, Romero L, Solano D (2019) Preservation of confidential information privacy and association rule hiding for data mining: a bibliometric review. Procedia Comput. Sci. 151:1219–1224. R
3. Douglas H (2018) retail—origin and meaning of retail by Online Etymology Dictionary, https://goo.gl/zzwvu2. Acceded 25 May 2018
4. Tovar L, Montoya J, Martelo R (2018) Sistema ecléctico de filtrado de información basado en inteligencia computacional para recomendación de atractivos turísticos del Caribe Colombiano. Recuperado el 11 de 07 de 2018, de. http://revistas.sena.edu.co/index.php/LOG/article/download/1521/1692
5. Viloria A, Viviana Robayo P (2016) Virtual network level of application composed IP networks connected with systems—(NETS Peer-to-Peer). Indian J Sci Technol 9(46). https://doi.org/10.17485/ijst/2016/v9i46/107376
6. Witten I, Frank E (2000) Data mining: practical machine learning tools with Java implementations. M. Kaufmann
7. LEY No 1352/88. http://www.impuestospy.com/Leyes/Ley%20135288.php; JlaweJ H, Kamber M (2002) Data mining: concepts and techniques. Simón Fraser University. Morgan Kaufmann Publishers
8. Moreno A, Redondo T (2016) Text analytics: the convergence of Big Data and artificial intelligence. Int. J Interact Multimed Artif Intell 2016
9. Silva J, Solano D, Fernández C, Romero L, Villa JV (2019) Privacy preserving, protection of personal data, and big data: a review of the colombia case. Procedia Comput Sci 151:1213–1218
10. Heaton J (2016) Comparing dataset characteristics that favor the Apriori, Eclat or FP- Growth frequent itemset mining algorithms. SoutheastCon. IEEE, pp 1–7
11. Bennet J, Lanning S (2007) The Netflix prizes. In: Proceedings of KDD cup and workshop, pp 3–6
12. Linyuan L, Medo M, Chi Ho Y, Yi-Cheng Z, Zi-Ke Z, Tao Z (2012) Recommender systems. Elsevier B.V, pp 1–49

13. López Puga JG (2007) Las redes bayesianas como herramientas de modelado en psicología. Anales de Psicología. Redalyc, 307–3016
14. Matich DJ (2001) Redes Neuronales: Conceptos básicos y aplicaciones. Cátedra de Informática Aplicada a la Ingeniería de Procesos-Orientación, Rosario
15. Lee J, Lee D, Lee YC, Hwang WS, Kim SW (2016) Improving the accuracy of top-n recommendation using a preference model. Inf Sci 348(1):290–304
16. Silva J, Varela N, Lezama OBP, Hernández PH, Ventura JM, de la Hoz B, Coronel LP (2019) Multi-dimension tensor factorization collaborative filtering recommendation for academic profiles. In: International symposium on neural networks Springer, Cham, pp 200–209
17. Kim J, Ale J (2004) Descubrimiento incremental de las reglas de asociación temporales. X Congreso Argentino de Ciencias de la Computacíon
18. Moine J, Gordillo S, Haedo A (2011) Análisis comparativo de metodologías para la gestión de proyectos de Minería de Datos. Congreso Argentino de Ciencias de la Computación
19. Sosa-Cabrera G, García-Torres M, Gómez S, Schaerer C, Divina F (2016) Understanding a version of multivariate symmetric uncertainty to assist in feature selection. In: Conference of computational interdisciplinary science)
20. Taneja A, Arora A (2018) Cross domain recommendation using multidimensional tensor factorization. Expert Syst Appl 92(1):304–316

Sleep Stage Classification Based on Ensemble Decision Tree Technique Using Single-Channel EEG

Pankaj Jadhav, Debabrata Datta, and Siddhartha Mukhopadhyay

Abstract The sleep stage classification can be used to study and characterize brain activities during sleep, which can be useful for the diagnosis and treatment of sleep-related diseases. Machine learning-based supervised classification methods for automatic sleep stage classification using EEG single-channel (Fpz-Cz and Pz-Oz) of PSG is proposed in the present study. The PSG dataset is obtained from the Physionet EDFx sleep Database. The 30 s epochs of single-channel EEG (either Fpz-Cz or Pz-Oz), are divided into 8 EEG subbands. From these subbands, a total of 96 features, i.e., twelve features per band are extracted. Ensemble-based classifier random forest is trained by using this labeled dataset of features for automatic sleep stage classification. The random forest classifier performance is compared for both channels, i.e., for healthy subjects and subjects having mild difficulty in sleeping.

Keywords EEG · Sleep stages · Machine learning · Random forest

1 Introduction

Sleep plays a vital role in the daily routine of human beings. Considering 60 years of average life span and 8 h of sleep a day, every human being spends an average of one-third of total life in sleep. Many diseases are caused due to lack of sleep. Hence, the study of sleep stages is essential for the diagnosis and treatment of sleep-related diseases.

Broadly, sleep stages are divided into wakefulness (W), non-rapid eye movements sleep (NREM), and rapid eye movement sleep (REM). Further, Rechtschaffen and Kales (R&K) guidelines [16] divide sleep into six stages such as W, NREM stage

P. Jadhav (✉) · D. Datta · S. Mukhopadhyay
Homi Bhabha National Institute, Mumbai, India
e-mail: pankaj.eln@gmail.com

D. Datta · S. Mukhopadhyay
Bhabha Atomic Research Centre, Mumbai, India

R. Patgiri et al. (eds.), *Proceedings of International Conference on Big Data, Machine Learning and Applications*, Lecture Notes in Networks and Systems 180,
https://doi.org/10.1007/978-981-33-4788-5_21

into four stages N1, N2, N3, N4, and REM. American Academy of sleep medicine (AASM) guidelines [3] divides sleep into only five stages. It combines stages N3 and N4 into stage N3 called slow-wave sleep or deep sleep.

Different sleep stages are distinguished based on signal amplitude, relative energy content at different frequency bands, presence or absence of EEG characteristics waves. Different bands of the EEG signals and their corresponding frequency ranges are given in Table 1. In Table 2, sleep stages with their EEG amplitude and characteristics waveforms are given. The properties of different sleep stages for visual classification are discussed further. The wake (W) stage consists of low amplitude mixed frequency EEG waveforms. It contains more than 50% frequencies in the alpha band (8–12 Hz). Stage N1 is the transition between wake and sleep. In the N1 stage, the person is responsive to external stimulus. It contains frequencies of alpha (8–12 Hz) and theta (4–8 Hz) band. Sleep spindles and K-complex characterize stage N2. Sleep spindles are sudden bursts of oscillatory brain activity. The sleep spindles consist of 12–15 Hz waves up to 50 μV that occurs for 0.5–1.5 s. K-complex (KC) is a high-amplitude biphasic wave which is composed of an initial negative sharp wave (deflection up) followed by a slow wave [25]. The stage N3 contains high voltage ($>$75 μV) and low frequencies in the delta (0–4 Hz) band of EEG. The stage REM consists of theta waves (4–8 Hz) and alpha waves (8–12 Hz), which is 1 2 Hz lower than the wake state. The REM state has quite a similar EEG waveform pattern to state N1.

For Automatic sleep stage classification (ASSC), the existing methods rely on polysomnographic (PSG) records, which include signals obtained from EEG (brain waves), EOG (eye movements), and EMG (muscle activity) recorded at the sleep laboratory. The classification of sleep stages is generally accomplished by visual inspection of these PSG signal epochs of 30 s each by sleep expert from EEG amplitude, frequency characteristics, and presence or absence of brainwaves such as sleep spindles and K-complexes. This process is tedious and requires expert knowledge. The automatic sleep stage classification is useful in the manual scoring of hypnogram and reduces the time and efforts of a sleep expert. Hence, an automatic classification of sleep stage is required.

Table 1 EEG frequency bands

Band	Frequency range (Hz)
Delta (δ)	0–4
Theta (Θ)	4–8
Alpha (α)	8–12
Sigma (σ)	12–15
Beta (β1)	15–22
Beta (β2)	22–30
Gamma (Γ1)	30–40
Gamma (Γ2)	40–49.5

Table 2 Sleep stages

Sleep stages	Amplitude (μV)	EEG characteristic waveforms
Wake (W)	<50	β (wake and active with eyes open), α(drowsy with eyes closed)
N1	50–100	θ
N2	50–150	θ, Sleep spindle, K-complex
N3	100–200	δ
REM	<50	β

For ASSC, pattern reorganization of recorded signals is important. Different machine learning and deep learning methods are proposed in the literature for ASSC based on pattern reorganization. ASSC is performed by extracting features from EEG in time domain [13], frequency domain , time-frequency domain [9, 14, 19]. The nonlinear features are also used for classification. The machine learning algorithms such as SVM [1], random forest [9], and artificial neural network [6, 20] are used for classification of sleep stages. The deep learning techniques, such as convolution neural network (CNN) extract features or patterns from data automatically [26]. The 1D CNN is used for sleep stage classification using the Fpz-Cz channel from EDFx database [31]. The 1D CNN to automatically extract features with bidirectional LSTM to use sleep stage transition rules is given in [30]. The features from the spectral power of frequency bands and Shannon entropy are used to train a neural network with LSTM to find the sequential pattern is given in [5]. Deep belief net is used for sleep stage classification using features from EEG, EMG, and EOG channels [22, 32]. In [21], features from eight EEG bands were used for sleep stage classification.

The challenges with existing techniques are as below. Existing techniques classify sleep stages using multiple signals such as EEG and EMG recorded at the sleep laboratory. The person under inspection is unable to sleep due to sensor mountings and the unfamiliar environment of a laboratory. Hence, there is a need to minimize the sensors to capture sleep information to classify sleep stages. Sleep stage classification using single-channel EEG reduces the processing of total PSG data and can be used for designing a wearable system for monitoring the sleep stage. In this paper, an automatic classification of sleep stages is presented according to AASM rules in five stages wake (W), N1, N2, N3, and REM using single-channel EEG. The various features have been calculated from EEG subbands used to classify sleep using a random forest classifier method. The results obtained by these methods have been compared.

The paper is organized as follows: Section 2.1 describes the dataset used for ASSC. In Sect. 2.2, preprocessing of EEG signal, feature extraction, ASSC methods, and classifier performance metrics are given. The classification results of sleep stages using a random forest algorithm are discussed in Sect. 3. Section 4 presents the conclusion.

2 Method

The visual sleep stage classification is done by dividing continuous PSG records into 30 sec epochs. Automatic sleep stage classification is a three-step process: EEG signal preprocessing, feature extraction, and classification using machine learning algorithm. The simplified block diagram of automatic sleep stage classification system is shown in Fig. 1.

2.1 Data

Sleep European Data Format (sleep-EDFx) dataset is used to study sleep stage classification [12, 18]. It consists of 61 PSG records with a corresponding hypnogram file. The 61 records consist of 39 SC∗ files (SC = Sleep Cassette) of 20 healthy subjects without any sleep medication recorded during two subsequent day-night each of 20 h. One subject's one-night recording is missing. Remaining 22 ST∗ files (ST = Sleep Telemetry) were recorded during two nights of 9 h of 22 subjects having mild difficulty in sleeping. Each polysomnogram contains a subject's whole night sleep recording of EEG (from Fpz-Cz and Pz-Oz electrode locations), EOG and submental chin EMG, and an event marker. The sampling frequency of EEG channels is 100 Hz. These PSG sleep recordings are manually classified in successive windows of 30 sec by sleep experts based on R&K guidelines. It classifies sleep stages into W (wake), R (REM), 1 (N1), 2 (N2), 3 (N3), 4 (N4), M (Movement time), and not scored epochs. Hypnogram files contain annotations of sleep type that corresponds to every 30 s polysomnographic records. The number of epochs corresponding to different sleep stages is given in Fig. 2 and they are equal in both EEG channels. It can be inferred

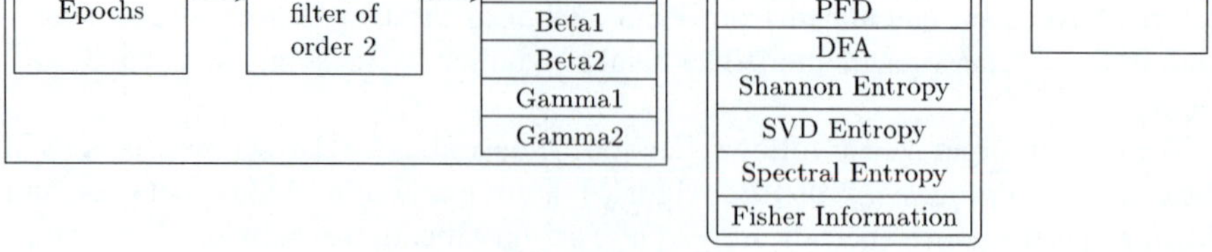

Fig. 1 Block diagram of sleep stage classification system

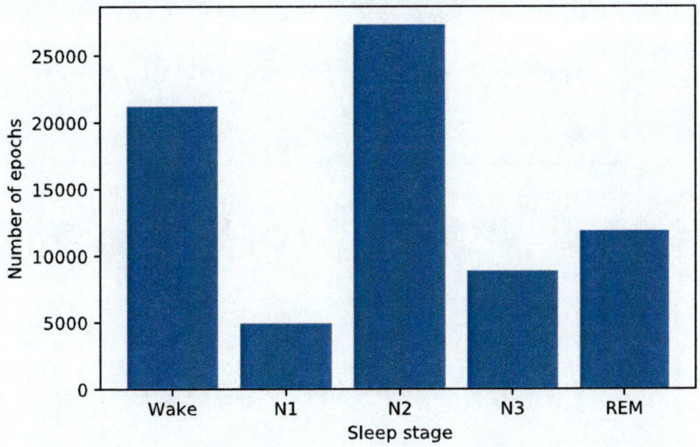

Fig. 2 Total number of epochs for sleep stages classification

from the figure that, class distribution of sleep stages is highly imbalanced, which is a severe concern in any biomedical signal classification.

2.2 *EEG Signal Preprocessing*

Noise and artifacts in the EEG signal are removed by using filtering [27] or by using wavelet techniques [23]. In this study, the EEG channels Fpz-Cz and Pz-Oz are divided into 30 sec epochs according to annotations given in the hypnogram. While making this dataset, (M) movement time and not scored epochs are ignored and sleep stage 3 and stage 4 are treated as stage 3 as per AASM guidelines. The various sleep stage waveforms are shown in Fig. 3. The waveforms are taken from the Fpz-Cz EEG channel. The epochs are further divided into eight frequency subbands as given in Table 1 by using a Butterworth bandpass filter of order 2.

2.3 *Feature Extraction*

Different features are extracted from EEG subbands that give information about the sleep state. The feature set includes features from the time domain, frequency domain, and nonlinear features.

1. Standard Deviation: It is a statistical feature that represents the distribution of data with respect to the mean, i.e., the spread of data. Standard deviation is the second moment in statistics and is defined as

$$\sigma_x = \sqrt{\frac{1}{N-1}\sum_{n=1}^{N}\left(x\left(n\right)-\bar{x}\right)^2} \tag{1}$$

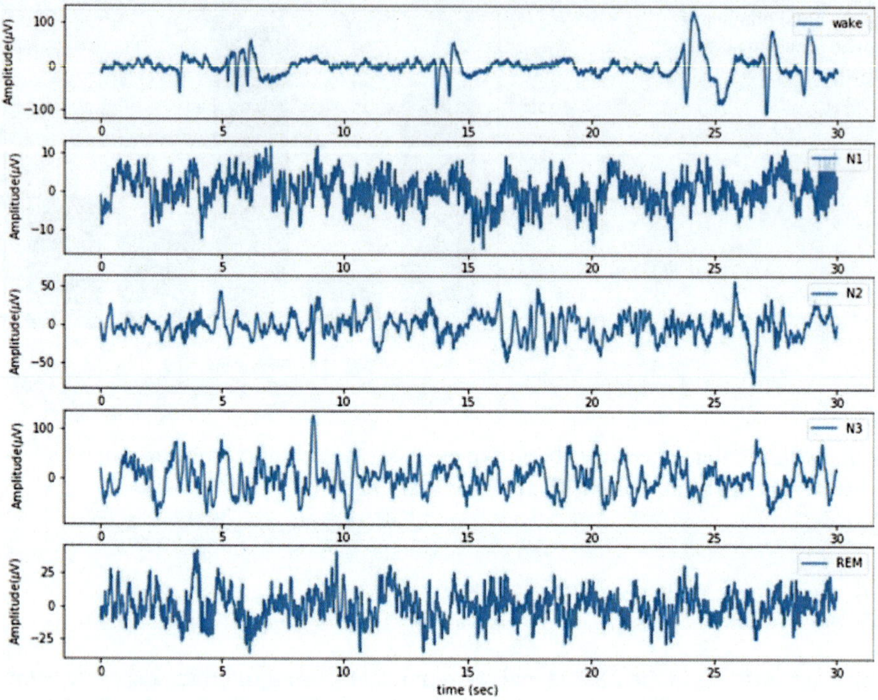

Fig. 3 Waveforms of different sleep stages from Fpz-Cz EEG channel

where $\bar{x}$ is mean of data given by

$$\bar{x} = \frac{1}{N}\sum_{n=1}^{N} x\,(n) \tag{2}$$

2. Skewness: Skewness gives the amount and direction of the skew of data. Skewness is the third moment in statistics and is defined as

$$S = \frac{1}{N}\sum_{n=1}^{N}\left[\frac{x_n - \bar{x}}{\sigma_x}\right]^3 \tag{3}$$

3. Kurtosis: Kurtosis gives how tall and sharp is the central peak of data. Coefficients of EEG have heavy-tail distribution. Kurtosis is the fourth moment in statistics and is defined as

$$k = \frac{1}{N}\sum_{n=1}^{N}\left[\frac{x_n - \bar{x}}{\sigma_x}\right]^4 \tag{4}$$

4. Hjorth Parameters: Hjorth parameters give statistical properties of signal in time domain [15]. Hjorth parameters are also called "normalized slope descriptors" because they can be defined using first and second derivatives. Hjorth activity (HA) is a measure of the mean power, i.e., variance representing the activity of the signal. Hjorth mobility (HM) is an estimate of the mean frequency. It is the ratio of, the standard deviation of the first derivative of the signal to the standard deviation of the signal. Hjorth complexity (HC) represents a change in frequency; the parameter compares the signal's similarity to a pure sine wave, value converges to 1 if the signal is more similar. For ASSC, the HM and HC features are considered in this study. Let x(n) is signal, and its first and second derivates are $x'(n)$ and $x''(n)$, respectively

$$\text{HA} = \text{var}(x) \tag{5}$$

$$\text{HM} = \frac{\sigma_{x'}}{\sigma_x} \tag{6}$$

$$\text{HC} = \frac{(\sigma_{x''}/\sigma_{x'})}{(\sigma_{x'}/\sigma_x)} \tag{7}$$

5. Maximum-Minimum Distance (MMD): The MMD feature is based on the Pythagoras theorem [2]. The distance d_k between maximum and minimum value in the sub-window is calculated using the Pythagoras theorem.

$$d_k = \sqrt{\Delta x_k{}^2 + \Delta y_k{}^2} \tag{8}$$

where, Δx_k and Δy_k represents difference in position and amplitude of maximum and minimum value in sub-window. The MMD is obtained by the addition of this distance (d_k) of all sub-window in epochs.

$$\text{MMD} = \sum_{k=1}^{N} |d_k| \tag{9}$$

6. Petrosian Fractal Dimension (PFD): The fractal dimension of the signal is a useful tool for transient detection in EEG and ECG analysis [7]. Petrosian fractal dimension is used for rapid computation of fractal dimension [11]. On the subtraction of adjacent samples, the signal is transformed into a binary sequence, which is represented by "+1" and "−1". PFD is computed by

$$\text{PFD} = \frac{\log_{10} N}{\log_{10} N + \log_{10}\left(\frac{N}{(N+0.4N_{\Delta})}\right)} \tag{10}$$

where N is the length of sequence and N_{Δ} is number of sign changes in a binary sequence.

7. Detrended Fluctuation Analysis (DFA): DFA is used for sleep stage classification into three classes wake, REM and NREM in [8]. The DFA of time series $\{x_1, x_{2,} x_{3,...,} x_N\}$ is computed using following steps [24]:

   – Remove the mean from time series data

$$\tilde{x}_i = x_i - \bar{x} \tag{11}$$

   – Integrate the time series data (apply cumulative sum)

$$y_i = \sum_{i=1}^{j} \tilde{x}_i \qquad j = 0, 1, 2, \ldots, N \tag{12}$$

   – Divide the time series into n equal windows of length L, $n = N\ L$
   – Fit the least square line to each window
   – Detrend each window with local trend (y_{ni})
   – The root mean square deviation of each window from local trend i.e fluctuation is calculated for different values of n

$$F(n) = \sqrt{\frac{1}{n}\sum_{i=1}^{n} [y_i - y_{ni}]^2} \tag{13}$$

   This process is repeated for different window lengths. The F(n) increases with window length n. The log-log plot of F(n) vs. n shows the presence of power-law scaling. The fluctuation is characterized by scaling exponent α, the slope of line relating $\log(F(n))$ to $\log(n)$.

8. Shannon Entropy: Entropy is a measure of uncertainty of the EEG signal. Shannon introduced entropy in the field of information theory to measure the amount of information contained in signal in [29]. Entropy measures the randomness or pattern of the EEG signal. It is defined as

$$H_{\text{shannon}} = \sum_{i=1}^{N} p(x_i) \log_2 \left(\frac{1}{p(x_i)}\right) \tag{14}$$

9. SVD Entropy: This entropy was proposed in [28], which makes use of both singular value decomposition (SVD) and Shannon entropy to measure complexity.

   – To compute SVD from time series $\{x_1, x_{2,} x_{3,...,} x_N\}$ it is converted to embedded matrix of size $(M \times m)$ with lag J in time series and embedded dimension m where $m = N - (m - 1)J$
   – The i-th row of embedded matrix is $x_i = [x_i, x_{i+1}, \ldots, x_{i+(m-1)J}]$
   – Singular value decomposition is performed on embedded matrix to obtain M singular values $\sigma_1, \sigma_2, \ldots, \sigma_M$

- These singular values are normalized $\bar{\sigma_i} = \frac{\sigma_i}{\sum_{j=1}^{M} \sigma_j}$
- The SVD entropy is calculed as

$$H_{svd} = \sum_{i=1}^{M} \bar{\sigma_i} \log_2 \left(\frac{1}{\bar{\sigma_i}} \right) \quad (15)$$

10. Fisher Information: Fisher information [17] is used to characterize non-stationarity in a complex process like EEG and other biomedical signals. The Fisher information is obtained using normalized singular values used in Eq. (15).

$$I = \sum_{i=1}^{M} \frac{(\bar{\sigma}_{i+1} - \bar{\sigma_i})^2}{\bar{\sigma_i}} \quad (16)$$

11. Spectral Entropy: Spectral entropy is calculated using power spectral density (PSD) of the EEG signal. Spectral entropy is calculated in the following steps:
   - Calculate FFT of signal $X(w_i)$
   - Calculate PSD of signal as

$$P(w_i) = \frac{1}{N} |X(w_i)|^2 \quad (17)$$

   - Normalize PSD so that it can be treated as probability density function

$$P(w_i) = \frac{P(w_i)}{\sum_{w_j=0}^{w_j=w_w/2} P(w_j)} \quad (18)$$

   - The power spectral entropy is calculated as

$$H_{\text{spectral}} = \sum_{w_i=0}^{w_i=w_s/2} P(w_i) \log_2 \frac{1}{P(w_i)} \quad (19)$$

2.4 Classification Using Random Forest Method

In the ensemble learning method, instead of learning a single (weak) classifier, learn many weak classifiers that are good at different parts of the input space. The output class is given by the average (or weighted average) of the vote of each classifier.

Random forest is an ensemble-based classification method proposed by Breiman [4]. It combines the classification results from multiple decision trees and classification output is given by majority voting. In this method, decision trees are built using CART (Classification And Regression Tree) without a pruning algorithm. These

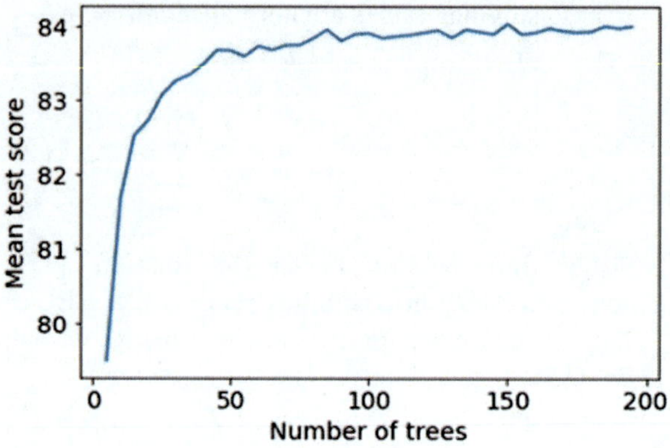

Fig. 4 Classification accuracy versus number of trees in random forest model

decision trees are formed independently using subsets of training dataset by sampling with replacement strategy (bootstrap sampling method). The input training set of different decision trees are selected as uncorrelated as possible. The additional randomness is added by selecting a subset of features randomly from these features for each decision. At each node of the decision tree, a subset of features is chosen randomly from total features. Then, nodes are split by using the feature having the highest Gini index. This strategy helps in adding randomness in the model to generate independent trees. The random forest is easily scalable to large datasets. It gives results by averaging the output from multiple decision trees. Hence, it avoids the overfitting problem and increases the classification accuracy. Correlation between trees is reduced by introducing randomness.

1. For b = 1, 2, 3, …, B
   (a) Draw a bootstrap dataset Z^*
   (b) Learn a tree $f_b(.)$ on Z^*, in particular select m (Typically take $m = \sqrt{p}$) features randomly out of p features as candidates before splitting
2. Classification: majority vote $f(x) = \frac{1}{B}\sum_{b=1}^{B} f_b(x)$

Figure 4 shows classification accuracy increases initially with the number of trees in the random forest, then it gets saturated near about 50 trees. The addition of further trees does not increase the accuracy significantly. Hence, 50 decision trees are used in this study. Features equal to the square root of the total number of features are used to generate each decision tree in the forest. Gini index is used to measure the quality of the node split while making the tree. Each node of tree is expanded until it contains less than two features. Thus, at least two features are required to split an internal node.

2.5 Performance Measure of Classifier

Classification accuracy, sensitivity, and specificity metrics show a better interpretation of performance measurement. Hence, to measure the performance of the sleep stage classifier model, sensitivity, specificity, and accuracy metrics are computed.

Sensitivity measures the proportion of positives that are correctly classified, i.e. the ratio of number of true positive instances to number of total positive instances.

$$\text{Sensitivity } (S_e) = \frac{TP}{(TP + FN)} \tag{20}$$

Specificity measures the proportion of negatives that are correctly classified, i.e. ratio of number of true negative instances to number of total negative instances.

$$\text{Specificity } \left(S_p\right) = \frac{TN}{(TN + FP)} \tag{21}$$

Accuracy is defined as ratio of number of correctly classified instances to all instances.

$$\text{Accuracy } (A_c) = \frac{(TP + TN)}{(TP + FP + TN + FN)} \tag{22}$$

where *TP*, *TN*, *FP*, and *FN* are true positive, true negative, false positive, and false negative, respectively.

3 Results and Discussion

The sleep stage classification is performed according to AASM guidelines into five sleep stages as wake, N1, N2, N3, and REM. The random forest method is used for classification by extracting the features from EEG subbands.

The performance of the classifier is evaluated using subject-wise K-fold cross-validation (CV). Each fold is data from one subject. In subject-wise K-fold CV data from one subject is used testing while remaining subjects (K-1) data is used for training classifier. The classifier results using a 20-fold COV (using SC subjects), 22-fold CV (using ST subjects), and 42-fold CV (using SC and ST subjects) are outlined. Classification results of Fpz-Cz and Pz-Oz models are summarized in the confusion matrix. The diagonal elements of the matrix represent the true positive case where the predicted and desired class is the same.

Table 3 shows the evaluation results for random forest classifier on healthy Caucasian subjects, which includes confusion matrix, sensitivity, specificity and accuracy of each class. The classifier model using the Fpz-Cz channel achieves overall sensitivity, specificity, and accuracy equal to 70.93%, 94.74%, 72.48%, respectively. The most accurately classified sleep stage is wake with 89.14% sensitivity. The classifier model using the Pz-Oz channel gives slightly less accurate results as compared to the Fpz-Cz channel. It provides overall sensitivity, specificity, and accuracy equal to 66.96%, 93.89%, 69.39%, respectively. For this channel also, sleep stage wake is the most accurately classified stage with 89.14% sensitivity.

Table 4 shows the evaluation results for random forest classifier on Caucasian subjects having mild difficulty falling asleep with temazepam medication, which includes confusion matrix, sensitivity, specificity, and accuracy of each class. The classifier model using the Fpz-Cz channel achieves overall sensitivity, specificity, and accuracy equal to 67.84%, 92.97%, 69.18%, respectively. The most accurately classified sleep stage is N2, with 89.14% sensitivity. The classifier model using the Pz-Oz channel gives slightly less accurate results as compared to the Fpz-Cz channel. It provides overall sensitivity, specificity and accuracy equal to 65.27%, 92.04%, 66.99%, respectively. For this channel also, sleep stage wake is the most accurately classified stage with 84.64% sensitivity.

Table 5 shows the evaluation results for random forest classifier on healthy Caucasian subjects having mild difficulty falling asleep with temazepam medication, which includes confusion matrix, sensitivity, specificity, and accuracy of each class. The classifier model using the Fpz-Cz channel achieves overall sensitivity, specificity, and accuracy equal to 70.61%, 94.31%, 71.69%, respectively. The most accurately classified sleep stage is wake with 88.79% sensitivity. The classifier model using the Pz-Oz channel gives slightly less accurate results as compared to the Fpz-Cz channel. It provides overall sensitivity, specificity, and accuracy equal to 66.59%, 93.44%, 68.66%, respectively. For this channel also, sleep stage wake is the most accurately classified stage with 86.61% sensitivity.

4 Conclusion

In this study, the random forest method is used to perform sleep stage classification. This study is used in the identification of brain function during sleep, as well as other clinical diagnoses. Visual sleep stage scoring done by experts varies from expert to expert, a human error occurs due to human eye and brain fatigue. The automatic classifier overcomes this limitation. The combination of different features that extract frequency, as well as amplitude variation of various sleep stages improves classification accuracy. Along with classifiers, preprocessing and feature extraction also plays an essential role in an accuracy improvement. The classifier model has good sensitivity and specificity. The proposed approach in this paper exhibits strong classification performance and can be used to assist sleep experts in manual sleep scoring. This analysis shows that Fpz-Cz gives higher accuracy than the Pz-Oz channel. The N1 and REM stages have similar characteristics, so it is difficult to distinguish these stages by the automatic classifier, as well as manual visual classification. Our method gives good sensitivity and specificity for N1 and REM stages. The comparison shows that the single-channel based sleep stage classification is accurate enough. The proposed approach in this paper exhibits strong classification performance and can be used to assist sleep experts in manual sleep scoring.

Table 3 Random forest performance results on healthy Caucasian subjects

	Automatic classifier score (Fpz-Cz)						
Expert score		Wake	N1	N2	N3	REM	Overall
	Wake	**7385**	395	144	48	313	8285
	N1	701	**527**	550	9	1017	2804
	N2	455	168	**15580**	493	1103	17799
	N3	138	1	788	**4765**	11	5703
	REM	539	383	954	4	**5837**	7717
Sn %		89.14	18.79	87.53	83.55	75.64	70.93
Sp %		94.61	97.6	90.06	98.49	92.93	94.74
Acc %		80.11	35.75	86.48	89.58	70.49	72.48

	Automatic classifier score (Pz-Oz)						
Expert score		Wake	N1	N2	N3	REM	Overall
	Wake	**7320**	382	208	12	363	8285
	N1	868	**307**	574	13	1042	2804
	N2	386	88	**15468**	722	1135	17799
	N3	82	3	1417	**4184**	17	5703
	REM	357	242	1305	6	**5807**	7717
Sn %		88.35	10.95	86.9	73.36	75.25	66.96
Sp %		95.02	98.19	85.7	97.94	92.61	93.89
Acc %		81.22	30.04	81.53	84.75	69.43	69.39

Table 4 Random forest performance results on Caucasian subjects having mild difficulty falling asleep with temazepam medication

	Automatic Classifier Score (Fpz-Cz)						
Expert Score		Wake	N1	N2	N3	REM	Overall
	Wake	**1970**	295	33	0	91	2389
	N1	389	**624**	427	2	602	2044
	N2	90	394	**8157**	543	309	9493
	N3	23	13	996	**2094**	19	3145
	REM	109	461	511	5	**3045**	4131
Sn %		82.46	30.53	85.93	66.58	73.71	67.84
Sp %		96.75	93.93	83.2	96.95	94.02	92.97
Acc %		76.33	34.92	80.57	79.2	74.89	69.18

	Automatic Classifier Score (Pz-Oz)						
Expert Score		Wake	N1	N2	N3	REM	Overall
	Wake	**2022**	295	35	0	37	2389
	N1	460	**577**	371	2	634	2044
	N2	105	368	**7862**	537	621	9493
	N3	14	26	1245	**1813**	47	3145
	REM	68	327	710	9	**3017**	4131
Sn %		84.64	28.23	82.82	57.65	73.03	65.27
Sp %		96.56	94.7	79.84	96.97	92.16	92.04
Acc %		75.76	36.22	76.91	76.79	69.26	66.99

Table 5 Random forest performance results on healthy Caucasian subjects having mild difficulty falling asleep with temazepam medication

	Automatic classifier score (Fpz-Cz)						
Expert Score		Wake	N1	N2	N3	REM	Overall
	Wake	**9477**	573	138	47	439	10674
	N1	1145	**1091**	1003	12	1597	4848
	N2	556	472	**23652**	1106	1506	27292
	N3	158	5	1713	**6946**	26	8848
	REM	678	794	1295	9	**9072**	11848
Sn %		88.79	22.5	86.66	78.5	76.57	70.61
Sp %		95.2	96.86	88.54	97.85	93.09	94.31
Acc %		78.88	37.17	85.08	85.54	71.77	71.69

	Automatic classifier score (Pz-Oz)						
Expert Score		Wake	N1	N2	N3	REM	Overall
	Wake	**9245**	707	274	16	432	10674
	N1	1410	**681**	904	20	1833	4848
	N2	433	376	**23551**	1200	1732	27292
	N3	96	20	2513	**6164**	55	8848
	REM	354	460	1981	10	**9043**	11848
Sn %		86.61	14.05	86.29	69.67	76.33	66.59
Sp %		95.66	97.34	84.34	97.72	92.16	93.44
Acc %		80.13	30.35	80.59	83.18	69.06	68.66

References

1. Aboalayon KA, Faezipour M (2014) Multi-class svm based on sleep stage identification using eeg signal. In: Healthcare innovation conference (HIC). IEEE. pp 181–184
2. Aboalayon KAI, Faezipour M, Almuhammadi WS, Moslehpour S (2016) Sleep stage classification using EEG signal analysis: a comprehensive survey and new investigation. Entropy 18(9):272
3. Berry RB, Brooks R, Gamaldo CE, Harding SM, Marcus C, Vaughn B (2012) The aasm manual for the scoring of sleep and associated events. Rules, terminology and technical specifications. Darien, Illinois, American Academy of Sleep Medicine
4. Breiman L (2001) Random forests. Mach Learn 45(1):5–32
5. Dong H, Supratak A, Pan W, Wu C, Matthews PM, Guo Y (2018) Mixed neural network approach for temporal sleep stage classification. IEEE Trans Neural Syst Rehabil Eng 26(2):324–333
6. Ebrahimi F, Mikaeili M, Estrada E, Nazeran H (2008) Automatic sleep stage classification based on eeg signals by using neural networks and wavelet packet coefficients. In: 30th annual international conference of the IEEE on engineering in medicine and biology society. EMBS 2008. IEEE, pp 1151–1154
7. Esteller R, Vachtsevanos G, Echauz J, Litt B (2001) A comparison of waveform fractal dimension algorithms. IEEE Trans Circuits Systems I: Fundam Theory Appl 48(2):177–183
8. Farag AF, El-Metwally SM, Morsy AAA (2013) Automated sleep staging using detrended fluctuation analysis of sleep EEG. In: Soft computing applications. Springer, pp 501–510
9. Fraiwan L, Lweesy K, Khasawneh N, Wenz H, Dickhaus H (2012) Automated sleep stage identification system based on time-frequency analysis of a single EEG channel and random forest classifier. Comput Methods Programs in Biomed 108(1):10–19
10. Friedman JH (2001) Greedy function approximation: a gradient boosting machine. Ann Stat 1189–1232
11. Goh C, Hamadicharef B, Henderson G, Ifeachor E (2005) Comparison of fractal dimension algorithms for the computation of eeg biomarkers for dementia. In: 2nd international conference on computational intelligence in medicine and healthcare (CIMED2005)
12. Goldberger AL, Amaral LA, Glass L, Hausdorff JM, Ivanov PC, Mark RG, Mietus JE, Moody GB, Peng CK, Stanley HE (2000) Physiobank, physiotoolkit, and physionet. Circulation 101(23):e215–e220
13. Hassan AR, Bashar SK, Bhuiyan MIH (2015) On the classification of sleep states by means of statistical and spectral features from single channel electroencephalogram. In: ICACCI. pp 2238–2243
14. Hassan AR, Bhuiyan MIH (2016) Automatic sleep scoring using statistical features in the EMD domain and ensemble methods. Biocybern Biomed Eng 36(1):248–255
15. Hjorth B (1970) EEG analysis based on time domain properties. Electroencephalogr Clin Neurophysiol 29(3):306–310
16. Hobson JA (1969) A manual of standardized terminology, techniques and scoring system for sleep stages of human subjects. In: Rechtschaffen A, Kales A (eds) Public health service, us government printing office. Washington, DC, p 58, Electroencephalography and clinical neurophysiology, vol 26, no 6, p 644
17. James CJ, Lowe D (2003) Extracting multisource brain activity from a single electromagnetic channel. Artif Intelli Med 28(1):89–104
18. Kemp B, Zwinderman AH, Tuk B, Kamphuisen HA, Oberye JJ (2000) Analysis of a sleep-dependent neuronal feedback loop: the slow-wave microcontinuity of the EEG. IEEE Trans Biomed Eng 47(9):1185–1194
19. Khalighi S, Sousa T, Pires G, Nunes U (2013) Automatic sleep staging: a computer assisted approach for optimal combination of features and polysomnographic channels. Expert Syst Appl 40(17):7046–7059

20. Liu Y, Yan L, Zeng B, Wang W (2010) Automatic sleep stage scoring using hilbert-huang transform with bp neural network. In: 2010 4th International Conference on bioinformatics and biomedical engineering (iCBBE). IEEE, pp 1–4
21. Memar P, Faradji F (2018) A novel multi-class EEG-based sleep stage classification system. IEEE Trans Neural Syst Rehabil Eng 26(1):84–95
22. Movahedi F, Coyle JL, Sejdić E (2018) Deep belief networks for electroencephalography: a review of recent contributions and future outlooks. IEEE J Biomed Health Inf 22(3):642–652
23. Oropesa E, Cycon HL, Jobert M (1999) Sleep stage classification using wavelet transform and neural network. International computer science institute
24. Peng CK, Havlin S, Stanley HE, Goldberger AL (1995) Quantification of scaling exponents and crossover phenomena in nonstationary heartbeat time series. Chaos: An Interdisc J Nonlinear Sci 5(1):82–87
25. Principe JC, Gala SK, Chang TG (1989) Sleep staging automaton based on the theory of evidence. IEEE Trans Biomed Eng 36(5):503–509
26. Rajpurkar P, Hannun AY, Haghpanahi M, Bourn C, Ng AY (2017) Cardiologist-level arrhythmia detection with convolutional neural networks. arXiv preprint arXiv:1707.01836
27. Rao R, Derakhshani R (2005) A comparison of EEG preprocessing methods using time delay neural networks. In: 2nd international IEEE EMBS conference on neural engineering, 2005. Conference proceedings. Citeseer, pp 262–264
28. Roberts SJ, Penny W, Rezek I (1999) Temporal and spatial complexity measures for electroencephalogram based brain-computer interfacing. Med Biol Eng Comput 37(1):93–98
29. Shin JW, Kim S.J (1949) A mathematical theory of communication. University of Ulinois Ress
30. Supratak A, Dong H, Wu C, Guo Y (2017) Deepsleepnet: a model for automatic sleep stage scoring based on raw single-channel EEG. IEEE Trans Neural Syst Rehabil Eng 25(11):1998–2008
31. Tsinalis O, Matthews PM, Guo Y, Zafeiriou S (2016) Automatic sleep stage scoring with single-channel EEG using convolutional neural networks. arXiv preprint arXiv:1610.01683
32. Zhang J, Wu Y, Bai J, Chen F (2016) Automatic sleep stage classification based on sparse deep belief net and combination of multiple classifiers. Trans Inst Meas Control 38(4):435–451
33. Zhu G, Li Y, Wen PP (2014) Analysis and classification of sleep stages based on difference visibility graphs from a single-channel EEG signal. REM 806:803

Author Index

R. Patgiri et al. (eds.), *Proceedings of International Conference on Big Data, Machine Learning and Applications*, Lecture Notes in Networks and Systems 180,
https://doi.org/10.1007/978-981-33-4788-5